U.S. Environmental Protection Agency
Office of Prevention, Pesticides, and Toxic Substances

Good Laboratory Practice Standards
Inspection Manual

Government Institutes, Inc.

Government Institutes, Inc.
4 Research Place, Suite 200, Rockville, Maryland 20850

Copyright © 1994 by Government Institutes. All rights reserved.
Published April 1994.

99 98 97 96 95 94   5 4 3 2 1

This manual was prepared by the U.S. Environmental Protection Agency, Office of Prevention, Pesticides, and Toxic Substances in September 1993. Government Institutes determined that it contained information of general interest, so we are reproducing this public domain information in order to serve those interested.

This manual is intended to provide guidance to EPA inspectors who conduct compliance inspections related to data submitted under Federal Insecticide, Fungicide, Rodenticide Act (FIFRA) Sections 3, 4, 5, 8, 18, and 24(c), and Toxic Substances Control Act (TSCA) Sections 4 and 5. This manual is intended as a supplement to other Office of Compliance Monitoring guidance materials pertaining to FIFRA and TSCA inspections and to environmental compliance inspections in general.

The publisher makes no representation of warranty, express or implied, as to the completeness, correctness, or utility of the information in this publication. In addition, the publisher assumes no liability whatsoever resulting from the use of or reliance upon the contents of this book.

EPA Disclaimer

Neither the United States Government nor its employees makes any warranty, expressed or implied, or assumes any legal liability for any third party's use of or the results of such use of any information, product, or process discussed in this document. Mention or illustration of company or trade names, or of commercial products, does not constitute endorsement by the U.S. Environmental Protection Agency.

ISBN: 0-86587-393-3

Printed in the United States of America

Table Of Contents

List Of Acronyms vii

1.0 Introduction 1-1

    1.1 Purpose Of This Manual 1-1
    1.2 Background On Good Laboratory Practice Standards 1-2
    1.3 FIFRA And TSCA GLP Standards 1-3
    1.4 GLP Enforcement Authority 1-3
        1.4.1 FIFRA Violations 1-3
        1.4.2 TSCA Violations 1-5
        1.4.3 GLP Violations 1-6
    1.5 Standards Of Professional Conduct 1-7

2.0 Pre-Inspection 2-1

    2.1 Introduction 2-1
    2.2 Determining The Scope Of The Inspection 2-1
        2.2.1 Compliance Reviews 2-1
        2.2.2 Study Audits 2-1
    2.3 Inspection Planning 2-2
    2.4 Reviewing EPA Information 2-3
    2.5 Providing Advance Notification 2-3
    2.5.1 Items Addressed in Advance Notification 2-4
    2.6 Inspection Team Coordination 2-5
        2.6.1 Inspection Plan 2-6
        2.6.2 Pre-Inspection Meeting with Team Members 2-6
        2.6.3 Preparation for Audits 2-6
    2.7 Gathering Inspection Documents And Equipment 2-7
    2.7.1 Types of Documents 2-7
        2.7.2 Inspection Equipment 2-9
    2.8 Confidential Business Information Considerations 2-9
        2.8.1 TSCA 2-9
        2.8.2 FIFRA 2-11

3.0 Inspection Procedures 3-1

    3.1 Introduction 3-1
    3.2 Inspection Authority 3-1
        3.2.1 Credentials 3-1
        3.2.2 Written Notice of Inspection 3-2
    3.3 Preliminary Steps For Conducting A GLP Inspection 3-2
    3.4 Consent To Enter And Inspect 3-5
        3.4.1 Procedures to Gain Consent 3-5
        3.4.2 Reluctance to Give Consent 3-7
        3.4.3 Conditional Consent 3-7
        3.4.4 Withdrawal of Consent 3-9
        3.4.5 Inspector Judgment 3-9

TABLE OF CONTENTS (Continued)

 3.5 Denial Of Consent To Enter And Inspect 3-10
 3.6 The Opening Conference 3-11
   3.6.1 Purpose 3-11
   3.6.2 Procedures 3-12
 3.7 The Closing Conference 3-15
   3.7.1 Purpose 3-15
   3.7.2 Pre-Closing Conference with Inspection Team 3-16
   3.7.3 Procedures 3-17
 3.8 Evidence Collection And Accountability 3-22
   3.8.1 Field Notebook 3-23
   3.8.2 Copies of Data, Records, Reports, and Correspondence ... 3-23
   3.8.3 Signed Statements 3-25
   3.8.4 Photographs and Observations 3-26
   3.8.5 Physical Sampling (Nondocumentary Samples) 3-27
   3.8.6 Maintenance of Inspection Materials 3-28
   3.8.7 FIFRA/TSCA Receipts 3-29
   3.8.8 Investigations Involving Alleged Criminal Activity 3-29

4.0 GLP Compliance Review 4-1

 4.1 Introduction 4-1
 4.2 Facility Compliance Review 4-1
 4.3 Organization And Personnel 4-2
   4.3.1 Personnel 4-2
   4.3.2 Testing Facility Management 4-3
   4.3.3 Study Director 4-4
   4.3.4 Quality Assurance Unit 4-6
 4.4 Facilities 4-7
   4.4.1 Test Facility, General 4-7
   4.4.2 Test System Care Facilities 4-8
   4.4.3 Test System Supply Facilities 4-8
   4.4.4 Facilities for Handling Test, Control, and Reference Substances 4-8
   4.4.5 Testing Facility Operations Areas 4-9
   4.4.6 Specimen and Data Storage Facilities 4-9
 4.5 Equipment 4-9
   4.5.1 Equipment Design 4-9
   4.5.2 Maintenance and Calibration 4-9
 4.6 Testing Facilities Operation 4-9
   4.6.1 Standard Operating Procedures 4-9
   4.6.2 Reagents and Solutions 4-10
   4.6.3 Animal and Other Test System Care 4-10
 4.7 Test, Control, And Reference Substances 4-11
   4.7.1 Test, Control, and Reference Substance Characterization 4-11
   4.7.2 Test, Control, and Reference Substance Handling 4-11
   4.7.3 Mixture of Substances with Carriers 4-12
 4.8 Protocol For And Conduct Of A Study 4-12
   4.8.1 Protocol 4-12
   4.8.2 Conduct of a Study 4-12

TABLE OF CONTENTS (Continued)

4.9	RECORDS AND REPORTS	4-13
	4.9.1 Reporting of Study Results	4-13
	4.9.2 Storage and Retrieval of Records	4-14
	4.9.3 Retention of Records	4-14
	4.9.4 Special Considerations for Field Sites	4-14
5.0	AUDIT PROCEDURES	5-1
	5.1 INTRODUCTION	5-1
	5.2 GLP COMPLIANCE	5-1
	5.3 DATA REVIEW	5-3
	5.4 COMMON DEFICIENCIES	5-5
	5.6 SOP REFERENCE LIST	5-5
6.0	POST-INSPECTION ACTIVITIES	6-1
	6.1 INTRODUCTION	6-1
	6.2 FOLLOWUP	6-1
	6.2.1 Followup Information from an Inspected Facility	6-2
	6.2.2 Followup Information from a Non-Inspected Facility	6-3
	6.3 THE INSPECTION REPORT	6-4
	6.4.1 Elements of the Inspection Report	6-5
	6.4.2 Inspection Checklists	6-6
	6.4.3 CBI Considerations	6-6
	6.4.4 Practical Tips for Report Preparation and Writing	6-6

APPENDICES

APPENDIX A - COMPLETED NOTIFICATION LETTER

APPENDIX B - INVESTIGATION REQUEST

APPENDIX C - EXAMPLE OF REQUEST FOR FURTHER INFORMATION LETTER

APPENDIX D - GLP FIFRA COMPLIANCE CHECKLIST

APPENDIX E - GLP TSCA COMPLIANCE CHECKLIST

List Of Tables

1-1 Major Differences Between FIFRA And TSCA 1-4

2-1 Types Of Equipment 2-10

List Of Figures

3-1 TSCA Notice Of Inspection 3-3

3-2 FIFRA Notice Of Inspection 3-4

3-3 TSCA Inspection Confidentiality Notice 3-13

3-4 TSCA Declaration Of Confidential Business Information Form 3-18

3-5 FIFRA Receipt For Samples 3-19

3-6 TSCA Receipt For Samples And Documents 3-20

LIST OF ACRONYMS

CAS	Chemical Abstracts Service
CBI	Confidential Business Information
CDO	Case Development Officer
CEO	Chief Executive Officer
CFR	Code of Federal Regulations
CIR	Comprehensive Inspection Report
DCO	Document Control Officer
ERP	Enforcement Response Policy
EPA	Environmental Protection Agency
FDA	Food and Drug Administration
FFDCA	Federal Food, Drug, and Cosmetic Act
FIFRA	Federal Insecticide, Fungicide, and Rodenticide Act
FOIA	Freedom of Information Act
GLP	Good Laboratory Practice
GRC	GLP Review Committee
IMD	Information Management Division
LDIAD	Laboratory Data Integrity Assurance Division
NEIC	National Enforcement Investigation Center
OCM	Office of Compliance Monitoring
OPP	Office of Pesticide Programs
OPPT	Office of Pollution Prevention and Toxics
QAU	Quality Assurance Unit
SOP	Standard Operating Procedure
SSB	Scientific Support Branch
TSCA	Toxic Substances Control Act
U.S.C.	United States Code

ACKNOWLEDGMENTS

The development of this manual was managed by John Helm of OCM, the EPA Work Assignment Manager; Fred Siegelman of OCM; and Diane Bradway of NEIC. Principal EPA contributors included Dawn Banks, Harry Day, Dean Hill, Steve Howie, Jack McCann, Randy Morris, as well as John Helm, Diane Bradway, and Fred Siegelman. Further assistance in the development of this manual was provided by Science Applications International Corporation (SAIC) in partial fulfillment of EPA Contract No. 68-D2-0157, Work Assignment 10.

1.0 Introduction

The U. S. Environmental Protection Agency (EPA) and the Food and Drug Administration (FDA) have promulgated Good Laboratory Practice (GLP) Standards Regulations to assure the quality of data submitted as part of the requirements of the Federal Insecticide, Fungicide, and Rodenticide Act (FIFRA), the Toxic Substances Control Act (TSCA), and the Federal Food, Drug, and Cosmetic Act (FFDCA). Data submitted to EPA under either FIFRA or TSCA requirements, must be prepared in accordance with the GLP standards. Based on an interagency agreement originated in 1978, FDA performs inspections for compliance at health effects laboratories that engage in testing for both FDA and EPA. EPA performs inspections to determine compliance and audit data at all other laboratories submitting to EPA under covered parts of FIFRA and TSCA.

1.1 Purpose Of This Manual

Congress delegated responsibility to EPA for promulgating and enforcing rules pursuant to FIFRA and TSCA. OCM is responsible for ensuring compliance with and enforcement of FIFRA and TSCA. A comprehensive compliance and enforcement program is necessary to assure that OCM meets its congressional mandate to protect health and the environment.

An integral part of OCM's compliance program is the compliance monitoring inspection, which is conducted to ascertain if regulated facilities are in compliance with FIFRA and TSCA regulatory requirements. This manual is intended to provide guidance to EPA inspectors who conduct compliance inspections related to data submitted under FIFRA Sections 3, 4, 5, 8, 18, and 24(c), and TSCA Sections 4 and 5. There are other OCM guidance materials pertaining to FIFRA and TSCA inspections and to environmental compliance inspections in general. This manual is intended as a supplement to those materials. Therefore, inspectors should refer to other documents as necessary when performing inspections to determine compliance with FIFRA and TSCA.

As the objectives of EPA, and OCM, change to reflect new priorities delegated by statutory, regulatory, or policy modifications, the methods, operations, and procedures employed by EPA in conducting GLP inspections may also change. The inspector should be aware that EPA may from time to time issue new Standard Operating Procedures (SOPs). These SOPs are the primary guidance documents developed by EPA to inform GLP inspectors of current policy and procedures. As such, all current SOPs will take precedence over the contents of this manual. In areas where this manual and newer SOPs differ, the directives of the SOPs will be followed.

This manual provides EPA inspectors with guidance in conducting GLP inspections under both FIFRA and TSCA. References to the applicable regulations will be for FIFRA (40 CFR Part 160) unless otherwise noted in the text. The corresponding TSCA regulations are found at 40 CFR Part 792.

1.2 BACKGROUND ON GOOD LABORATORY PRACTICE STANDARDS

The GLP regulations were developed to address problems found with data submitted to EPA and FDA. Investigations by these agencies in the mid-1970s revealed that some studies had not been conducted in accordance with commonly accepted laboratory practices. Some studies had been conducted so poorly that the resulting data could not be relied upon in EPA's regulatory decisionmaking process. As a result, the FIFRA and TSCA GLP standards developed by EPA and FDA specify minimum practices and procedures that must be followed to assure the quality and integrity of data submitted to EPA in support of applications for research or marketing permits for pesticide products in the case of FIFRA and for studies relating to health effects, environmental effects, and chemical fate in the case of TSCA.

The initial FIFRA and TSCA GLP standards were proposed in 1979, and final regulations became effective in May 1984, several years after FDA promulgated GLP standards under its statute. The 1984 regulations did not address all studies, although EPA recognized the need to do this at some point. EPA felt it needed more experience with some of the nonhealth effects studies in order to write GLP regulations applicable to them. It is important to note that while the 1984 GLP regulations specified a limited number and type of studies, the FIFRA Section 8 recordkeeping regulations [40 CFR 169.2(k)], which require that the original raw data supporting all studies be kept for the life of the pesticide's registration, did not. Thus, while FIFRA GLP requirements initially applied to a limited range of studies (i.e., health effects studies), EPA's books and records requirements encompassed all types of studies submitted to support registrations.

In August 1989, EPA published its final revised FIFRA and TSCA GLP rules, which expanded the coverage of the GLP standards to nearly all studies submitted to EPA under FIFRA or TSCA. The revisions also reflect changes made by FDA to its GLP regulations in 1987. EPA's revisions were made to assure some degree of consistency among the three rules. This was done to minimize the regulatory burden on laboratories that may conduct studies under all three statutes.

The revised TSCA GLP standards became effective on September 18, 1989, while the revised FIFRA GLP standards became effective on October 16, 1989. The FIFRA GLP standards Regulations are applicable to virtually any study required to be performed for submission to EPA in support of an application for a pesticide marketing or research permit. The TSCA GLP Standards regulation applies to any study required to be conducted to determine the effects of any chemical substance that is manufactured, distributed, processed, used, or disposed of within the jurisdiction of the United States, on human health and the environment. The purpose of the GLP standards is to provide some assurance that studies are conducted with certain safeguards in place regarding data quality and integrity and that the raw data, records, and reports will allow a study's reconstruction.

1.3 FIFRA AND TSCA GLP STANDARDS

EPA, through the Office of Compliance Monitoring (OCM), conducts GLP inspections under both TSCA (40 CFR Part 792) and FIFRA (40 CFR Part 160) regulations. In instances where FIFRA and TSCA differ, those differences will be cited in the ensuing sections of this manual. The two major differences between FIFRA and TSCA involve the scope of GLP applicability and the retention of records. These differences are shown in Table 1-1.

1.4 GLP ENFORCEMENT AUTHORITY

Violations of the GLP standards may be discovered in a number of ways. Generally, GLP violations are likely to be found (1) during laboratory inspections conducted by EPA and FDA, (2) during EPA's review of data submissions, or (3) through investigations of tips or complaints.

1.4.1 FIFRA Violations

Under FIFRA Section 12, it is unlawful for any person:

- To refuse to prepare, maintain, or submit any records required by or under Section 5, 7, 8, 11, or 19 [FIFRA §12(a)(2)(B)(i)].

- To refuse to submit any reports required by or under Section 5, 6, 7, 8, 11, or 19 [FIFRA §12(a)(2)(B)(ii)].

   Under Section 6 of FIFRA, registrants are required to submit any "additional factual information regarding unreasonable adverse effects on the environment..." [§6(a)(2)]. Failure to submit information required under this section is a violation of FIFRA §12(a)(2)(B)(ii).

- To knowingly falsify all or part of any application for registration, application for experimental use permit, any records required to be maintained pursuant to FIFRA, any report filed under FIFRA, or any information marked as confidential and submitted to the Administrator under any provision of FIFRA [FIFRA §12(a)(2)(M)].

- To falsify all or part of any information relating to the testing of any pesticide (or any ingredient, metabolite, or degradation product thereof), including the nature of any protocol, procedure, substance, organism, or equipment used, observation made, or conclusion or opinion formed, submitted to the Administrator, or that the person knows will be furnished to the Administrator or will become part of any records required to be maintained by FIFRA [FIFRA §12(a)(2)(Q)].

- To submit to the Administrator data known to be false in support of a registration [FIFRA §12(a)(2)(R)].

TABLE 1-1. MAJOR DIFFERENCES BETWEEN FIFRA AND TSCA GLP REGULATIONS	
FIFRA	**TSCA**
SCOPE OF DATA SUBJECT TO INSPECTION	
40 CFR §160.1(a): "... studies that support or are intended to **support applications for research or marketing permits** for pesticide products regulated by the EPA. This part is intended to assure the quality and integrity of data submitted pursuant to §§3, 4, 5, 8, 18, and 24(c) of FIFRA and §§408 and 409 of the FFDCA." **(b)**: "This part applies to any study ... of this section which any person conducts, initiates, or supports on or after 10/16/89."	**40 CFR §792.1**: "... studies relating **to health effects, environmental effects, and chemical fate testing**. This part is intended to assure the quality and integrity of data submitted pursuant to testing consent agreements and test rules issued under §4 of TSCA." **(b)**: "This part applies to any study ... of this section which any person conducts, initiates, or supports on or after 09/18/89." **(c)**: "It is EPA's policy that all data developed under §5 of TSCA be in accordance with this part. If data are not ... EPA will consider such data insufficient to evaluate the health and environmental effects of the chemical substances ..."
FIFRA	**TSCA**
RETENTION OF RECORDS	
40 CFR §160.195: "... documentation records, raw data, and specimens pertaining to a study and required to be retained by this part shall be retained in the archive(s) for whichever of the following periods is longest: (1) In the case of any study used to support an application for a research or marketing permit approved by EPA, **the period during which the sponsor holds any research or marketing permit** to which the study is pertinent. (2) A period of **at least 5 years** following the date on which the results of the study are submitted to the EPA in support of an application for a research or other marketing permit. (3) In other situations ... a period of **at least 2 years** following the date on which the study is completed, terminated, or discontinued."	**40 CFR §792.195(b)(1)**: "... documentation records, raw data, and specimens pertaining to a study and required to be retained by this part shall be retained in the archive(s) for a period of **at least ten (10) years**..."

FIFRA provides EPA with the authority to issue Notices of Warning for violations and to assess administrative civil penalties of up to $5,000 per violation. Any person who knowingly violates FIFRA may also incur criminal penalties of up to $50,000 or 1 year in jail. Falsification of submissions or records may be the basis for a criminal referral under the United States Code, Chapter 47 (18 U.S.C. §1001). Actions may be taken against the registrant, the laboratory, or individuals for falsifying the certification statements, records, or reports. Generally, EPA will examine the specific case to determine who is most appropriately subject to an enforcement action.

In addition to enforcement actions taken under FIFRA, failure to comply with all applicable sections may be used as the basis for rejection of studies. Where a study is not conducted in accordance with GLP standards, EPA may refuse to consider the data as meeting regulatory requirements. GLP deviations may also be the basis for the cancellation or suspension of a registration, modification of the research or marketing permit, or denial of an application for such a permit.

1.4.2 TSCA Violations

Under TSCA, it is unlawful for any person:

- To fail or refuse to comply with any rule promulgated or order issued under Section 4 of TSCA [TSCA §15(1)]

- To fail or refuse to (1) establish or maintain records; (2) submit reports, notices, or other information; or (3) permit access to or copying of records, as required by TSCA or a rule thereunder [TSCA §15(2)]

- To fail or refuse to permit entry or inspection as required by Section 11 of TSCA [TSCA §15(4)].

TSCA allows EPA to assess civil and administrative penalties of up to $25,000 per violation per day. Any person who knowingly violates any portion of TSCA may also be liable for criminal penalties of up to $25,000 per violation per day and/or up to 1 year in jail.

If the Administrator determines that a testing facility did not comply with any part of the GLP regulations, data submitted may be determined to be unreliable for the purposes of showing that a chemical substance or mixture does not present a risk of injury to health or the environment. If data submitted to EPA to fulfill a testing consent agreement or a test rule issued under Section 4 of TSCA are not developed in accordance with the applicable GLP regulations, EPA may require the sponsor of that testing consent agreement or a test rule data submission to develop data in accordance with the GLP standards before accepting those data [40 CFR § 792.17(c)]. In addition, it is EPA policy that all data developed under Section 5 of TSCA be in accordance with GLP regulations. EPA considers any Section 5 data submitted that was not in accord with the GLP regulations to be insufficient to evaluate the health and environmental effects of the chemical substance unless the submitter provides additional information demonstrating that the data are reliable and adequate.

1.4.3 GLP Violations

EPA has developed several enforcement response policy (ERP) documents to guide case development officers (CDOs) in determining the seriousness of GLP violations and the appropriate penalties to be assessed for those violations. The FIFRA Enforcement Response Policy, dated July 2, 1990, contains the EPA policy regarding penalty determinations for FIFRA violations. This ERP includes tables that list the gravity levels for each violation of FIFRA and matrices for determining appropriate penalty amounts based on the gravity, size of business, history of noncompliance, culpability of the violator, and harm to human health or the environment. The ERP applies to any violations of FIFRA, including those concerning GLP regulations. Specific FIFRA charges given in the ERP are in reference to Section 12(a)(2)(M), for knowing falsification of reports submitted to EPA; Section 12(a)(2)(Q), for falsifying information related to testing; and Section 12(a)(2)(R), for submission of data known to be false.

The FIFRA ERP addresses GLP violations in the following manner: a high-level GLP violation has a maximum civil penalty of $5,000 per violation; a middle-level GLP violation has a maximum civil penalty of $4,000 per violation; and a low-level GLP violation has a maximum civil penalty of $3,000 per violation.

The GLP ERP supplement to the FIFRA ERP was released on September 30, 1991. This document provides guidance regarding certain enforcement policy issues, including multiple GLP violations and liability for GLP violations. It also provides guidance regarding which violations are considered to be high-, middle-, or low-level violations.

EPA developed an ERP for TSCA GLPs in 1985. The TSCA ERP outlines both the levels of action, and the penalty amounts that can be imposed where violations have been discovered. The TSCA ERP for GLPs is applicable to those studies used to obtain data for TSCA section 4 hazard evaluations, TSCA section 5 data submissions, and negotiated testing agreements. The TSCA ERP identifies four response levels for GLP violations, including:

- Notice of Noncompliance (NONs) - NONs are the most common response to a TSCA GLP violation. They are used for minor, technical or form violations. They are not used for a substantive violation, or for any repeat offenses under TSCA section 4.

- Civil Administrative Penalties (CAPs) - CAPs are appropriate when one or more violations, considered together or separately, have the potential to affect the reliability and accuracy of the data.

- Criminal Sanctions - Criminal sanctions are used in serious cases of misconduct. The factors used to determine whether or not to proceed with a criminal prosecution include whether there was "guilty knowledge" or intent on the part of the responsible party, or where violations are "knowingly or willfully" committed (i.e., falsifying material data, intentional concealment of results through omission or selective reporting).

INTRODUCTION

- Study Invalidation - Under 40 CFR 792.17, EPA may determine that the submitted data was not collected in accordance with the applicable GLPs and may be unreliable for the purposes of indicating that the chemical in question is not expected to pose an unreasonable risk to human health or the environment. If the data is submitted as part of a TSCA section 4 study, EPA may require a new study to be conducted. For studies submitted under §5 or a negotiated testing agreement that do not conform to GLPs, EPA may consider the data insufficient to permit a reasoned evaluation of the environmental and health effects of a chemical substance.

Similar to the FIFRA GLP ERP, a penalty matrix was created to aid CDOs in determining the amount of fines to be leveled for CAPs. The ERP also delineates adjusting factors that may be taken into consideration when determining the appropriate response level to the GLP violation.

Inspectors are encouraged to review the FIFRA and TSCA GLP ERPs to familiarize themselves with the types and levels of GLP violations EPA has established.

1.5 STANDARDS OF PROFESSIONAL CONDUCT

Through many years of inspection experience, EPA has developed procedures and requirements that assure ethical action on the part of its inspectors. These ethics have been established to protect the individual, the Agency, and industry as well. Because inspectors act as officers of the United States Government, they should perform their duties with the highest degree of honesty and integrity. In addition, they are expected to conduct themselves in a manner that will reflect favorably on themselves and the Agency. As such, the following rules of ethics should be adhered to at all times:

- All investigations shall be conducted within the framework of the United States Constitution and with due consideration for individual rights, regardless of race, sex, creed, or national origin.

- The inspector shall uphold the Constitution, laws, and regulations of the United States and all governments therein and never be a party to their evasion.

- The inspector shall never use any information obtained confidentially in the performance of governmental duties as a means for making a private profit.

- Any act (or failure to act) that might be construed as being motivated by personal or private gain (conflict of interest) should be avoided.

- The inspector shall never discriminate by dispensing special favors or privileges to anyone, whether for renumeration or not; and never accept, for him/herself or his/her family, favors or benefits under any circumstances.

- Facts of an investigation are to be developed and reported completely, objectively, and accurately.

- The inspector shall make no promises of any kind; government employees cannot bind government enforcement.

- The inspector shall continually attempt to improve professional knowledge and technical skill in the investigative field.

2.0 PRE-INSPECTION PROCEDURES

2.1 INTRODUCTION

This chapter contains guidance on conducting pre-inspection activities pertaining to FIFRA and TSCA GLP inspections. A good inspection begins with planning, which should commence well before the inspector visits the subject facility. Planning is the process during which the inspector identifies all the required activities to be completed during the inspection process. These activities include obtaining records before the inspection, conducting the inspection, follow up, and writing the inspection report.

This chapter describes the planning process that should take place prior to any GLP inspection. The basic elements of inspection planning are determining the scope of the inspection (Section 2.2); inspection planning (Section 2.3); reviewing EPA information (Section 2.4); providing advance notification of the inspection to the facility (Section 2.5); coordinating the inspection team (Section 2.6); and gathering inspection documents and equipment (Section 2.7). Section 2.8 discusses Confidential Business Information (CBI) issues.

2.2 DETERMINING THE SCOPE OF THE INSPECTION

The first step in the inspection planning process is determining what type of inspection will be conducted. GLP inspections usually involve both a compliance review and one or more study audits. In some cases, only a study audit will be conducted.

2.2.1 Compliance Reviews

A compliance review is used to obtain a "snapshot in time" at a testing facility (i.e., to determine compliance status at the time the inspection takes place). During a compliance review, the inspection team may identify an ongoing or recently submitted study and evaluate to what extent the study is being conducted in accordance with GLP regulations. Practices evaluated may include any of the items or activities required in the GLP regulations such as use and maintenance of SOPs, data recording, handling of test systems, or other operations. A detailed discussion of all elements of a compliance review is presented in Chapter 4.

2.2.2 Study Audits

The purpose of a study audit is to determine whether the testing facility being inspected has documented the raw data necessary to support conclusions previously submitted to EPA and whether the GLP Standards were followed. During a study audit, auditors conduct an overall review of raw data, verify the accuracy of the data, and examine the submitted study report to determine whether the data were collected following the proper procedures required by the GLP regulations. A discussion of the aspects of study audits is presented in Chapter 5 and in the relevant SOPs.

2.3 INSPECTION PLANNING

Planning includes conducting a thorough review, prior to the inspection, of EPA records and other information pertaining to the facility to be inspected. This will save time because familiarity with the operation, history, and compliance status of the subject facility decreases the need for more extensive discussion of these areas during the limited time typically allotted to an onsite visit. In addition, planning promotes a better relationship with the regulated community because the inspector will be better able to answer questions concerning the application of GLP requirements to a particular type of facility.

Proper planning also enhances the facility personnel's confidence in the EPA inspector and aids in establishing good relationships with facility representatives. An inspector that knows what s/he wants, how to proceed, what to accomplish, and who articulates such goals to the facility personnel will appear well organized and in control. Such an appearance indicates to the facility personnel that the inspector is a professional and is concerned not only with using his/her own time effectively, but also the time, energy, and resources of the facility as well.

Another benefit of planning is that it enhances the inspector's ability to identify and document potential violations and thus provide more time to collect necessary data to assist CDOs in their subsequent compliance and enforcement activities. Planning an inspection will result in a more efficient and productive inspection overall.

The objectives of inspection planning are to:

- Understand the objectives of the inspection.

- Understand applicable GLP regulations.

- Be well-versed in the policies and procedures governing GLP inspections and the SOPs of the Laboratory Data Integrity Assurance Division (LDIAD).

- Obtain the proper equipment, material, and documents and/or forms for conducting the inspection.

- Be prepared to collect and record documentary, and if necessary, nondocumentary samples.

Once a facility has been selected for inspection, proper planning should assure the following:

- A properly focused inspection

- A systematic framework for comparing a facility's operating practices against applicable GLP regulations

- Use of the most efficient and effective approach for conducting the inspection, given the available personnel and funding

- Clearly established task assignments in the field for each member of the inspection team.

2.4 Reviewing EPA Information

The inspector's responsibilities are initiated by the receipt of the Investigation Request letter from the Director, LDIAD. For a study audit, copies of all study reports to be audited will be forwarded by the Scientific Support Branch (SSB) to the inspector and also to the auditor(s) when appropriate in a timely fashion (ideally at least 4 weeks prior to the start of the inspection). Members of the audit team will be provided with copies of pertinent study reports at the same time. The inspector should request any additional information that s/he will find useful in preparing for the inspection from a variety of Regional and Headquarters personnel. Such information may include (but is not limited to):

- Copies of previous inspection report(s)

- General facility information

- Correspondence with facility personnel

- Discussion with appropriate program staff, such as the Office of Pesticide Programs (OPP) reviewers

- Available study review documents from OPP or the Office of Pollution Prevention and Toxics (OPPT)

- Relevant program documents, such as FIFRA guidelines or TSCA test rules, and the SOPs of LDIAD.

2.5 Providing Advance Notification

EPA is not required by law to provide advance notice of an inspection. However, OCM has adopted a policy of providing such notification, based on the circumstances of the particular inspection and facility. It is up to the discretion of the SSB Branch Chief to decide whether to provide advance notification. Because of the highly technical nature of these inspections, the sensitivity of the information involved, and the need to assure that appropriate personnel and records are available for inspection, the testing facility, in most cases, is notified in advance that an inspection is planned. (For inspections involving certain well-documented complaints or tips, prior notification may not be given.)

Approximately 2 weeks before the scheduled inspection, the Chief of SSB (or another designated person) will contact the responsible management official at the facility and notify that person of the scheduled inspection. The initial telephone notification will be confirmed by a notification letter (see Appendix A).

CHAPTER TWO

The inspector will receive a copy of the notification letter. At the same time, the inspector will receive an investigation request (see Appendix B), which gives the name and telephone number of the facility contact person and of each inspection team member. This request also will provide other information necessary to the planning and conduct of the inspection.

Once the facility has been notified by SSB that an inspection will be conducted, the primary responsibility for the conduct of the inspection passes to the inspector. Any further communications with the facility personnel should be made by the inspector. The inspector should keep SSB and supervisory personnel apprised of the status of the inspection and should consult with them on any substantive issues that may arise or changes that may be required.

The potential advantages and disadvantages associated with providing advance notification are as follows:

- Potential advantages

    - The facility will have the necessary documents, records, or personnel available for the inspector, saving valuable time on site and requiring less time during followup stages of the inspection.

    - The facility personnel appreciate advance notification so that their regular operations are not interrupted, thereby fostering a cooperative relationship with EPA.

- Potential disadvantages

    - The inspector may not have the opportunity to view the facility under normal operating conditions because facility personnel, with advance notification, could tailor operations to fit preconceived notions of what the inspector may want to see.

    - At a facility suspected of violating GLP requirement(s), the company officials might conceal, alter, or destroy evidence confirming the violation(s) after receiving advance notification.

2.5.1 Items Addressed in Advance Notification

If advance notification is provided, the inspector should make note of it in the inspection report. Specific objectives of advance notification include the following:

- Identifying the inspector

- Scheduling the inspection (including establishing time of arrival)

- Obtaining verbal agreement to allow the inspection team to enter

- Determining the appropriate site(s) for the inspection, including identifying the location of necessary records, as specified in the inspection plan

- Ensuring that personnel are available to accompany EPA inspectors during the inspection

- Ensuring that someone onsite will be able to make claims of FIFRA/TSCA CBI

- Encouraging the facility and sponsor to have all records transferred to the inspection site before the inspection

- Obtaining directions to the facility

- Discussing problems, concerns, or questions relative to the inspection or studies to be audited or any other issues

- Ensuring there is photocopying capability and, if not, what document reproduction options are available to the inspection team.

When the facility has not previously been inspected for GLP standards compliance, the inspector should be certain that facility personnel are aware of what is involved in such an inspection, what records should be made available, what personnel should be present, etc. If the facility representative contacted does not cooperate, the inspector's supervisor and the Chief, SSB at EPA Headquarters should be consulted for instructions on how to proceed.

2.6 Inspection Team Coordination

As soon as the identity of the inspection team is known, the inspector should contact each person and begin planning the conduct of the inspection. As early as possible the inspector should:

- Coordinate travel plans, including the hotel to be used by the team, times of arrival of team members, means of transfer from the airport to the hotel, and provision for one or more rental cars of suitable size to accommodate the team.

- Ascertain that each team member is aware of the dates of the inspection, especially the date and time that s/he will be required to be available for a pre-inspection team meeting, and the expected date and time for the conclusion of the inspection. Assure that each team member is aware of the proper attire for the inspection and that has been briefed on appropriate safety procedures. The inspector should not underestimate the time needed to conduct the inspection.

- Confirm that those individuals who will be conducting the study audit portion of the inspection are aware of the studies (or portions of studies) to be audited. This is especially important for large or complex studies where more than one auditor will be reviewing data. If an auditor will be expected to assist with portions of the GLP compliance review of the facility, the inspector should discuss this with him/her during these early planning stages. Each member of the inspection team should be aware of his/her responsibilities during the compliance review and/or audit.

The inspector should also arrange to provide copies of applicable LDIAD SOPs to auditors who do not already have these documents. In addition, the inspector may need to assure that the inspection team is aware of proper procedures for receiving and handling CBI, especially as related to inspections conducted under TSCA. (TSCA CBI requirements are more stringent than those found under FIFRA, and the procedures for handling CBI under TSCA are more complicated.) The inspector may also need to coordinate with representatives of the EPA Regional offices, FDA, State agencies, contract organizations, or foreign governments. S/he should determine the level of experience of each auditor in conducting compliance reviews and/or study audits for studies conducted under GLP regulations. The inspector may need to provide guidance to less experienced auditors, both before and during the inspection.

2.6.1 Inspection Plan

Prior to the inspection, the inspector should prepare a plan for the inspection. S/he should remain flexible about this plan, since circumstances encountered at the facility may require last-minute changes. However, these can usually be minimized by adequate pre-inspection communication with SSB, the team members, and facility personnel. The inspection plan should include details such as date and time for the pre-inspection meeting with team members, date and time of arrival at the facility, name of facility contact person, assignment of responsibilities for GLP compliance review and study audits, and proposed timetable for accomplishing the compliance review and/or study audits. The inspector should assure that, if necessary, team members are cleared to handle FIFRA or TSCA CBI.

2.6.2 Pre-Inspection Meeting with Team Members

The pre-inspection meeting between the inspector and the team members usually occurs once the team has assembled just prior to the start of the inspection. The meeting provides an opportunity for the team members to get acquainted with each other and to attend to any last-minute details. It allows the inspector to verify that each team member is aware of his/her assignments and responsibilities and that each member understands the principles of adequate documentation and evidence gathering. During the meeting, the inspector should discuss the schedule and format for report preparation and assure that each team member has copies of the LDIAD SOPs for report preparation.

2.6.3 Preparation for Audits

The study auditor(s) also need adequate preparation. As early as possible in the audit planning, the auditor should receive and review the study(ies) to be audited to become familiar with the technical, management, and GLP aspects of the study. The auditor should also review the applicable GLP standards and the test rule or FIFRA guidelines for the study to be audited. During this review, discrepancies, deficiencies, and potential problem areas can often be identified and a plan developed for approaching the audit. The auditor may develop a number of questions to be discussed with study personnel.

The auditing preparation is, in part, used to determine if the studies were required to have been conducted in accordance with the GLP regulations. Often the inspector will ask the auditor to make that determination, particularly in those areas directly involving the auditor's field of expertise. If not already familiar with the FIFRA and TSCA GLP regulations (40 CFR Parts 160 and 792, respectively), the auditor should review these regulations and discuss any questions with the inspector.

Prior to the inspection, the auditor should review all pertinent LDIAD SOPs for conducting study audit(s) of the specific study types. These documents provide guidance and a standard procedure for conducting the various aspects of the GLP inspection. (For a complete index of SOPs, See Section 5.6.)

The auditor should always discuss any uncertainties about an upcoming inspection with the inspector. The inspector should make it clear that any problems encountered during the audit should be brought to the inspector's attention. The auditor should not confront facility personnel over any outstanding issues.

2.7 GATHERING INSPECTION DOCUMENTS AND EQUIPMENT

In addition to preparing the written inspection plan and reviewing EPA records prior to conducting the inspection, the inspector should gather and prepare the necessary documents and equipment to be used during the inspection.

2.7.1 Types of Documents

No single list of documents and equipment can be appropriate for all inspections. The list provided below is intended for guidance purposes only. The inspector's experience in the field and information obtained during pre-inspection planning should assist in preparing lists tailored to specific inspection sites and needs. Specific needs will be determined by the requirements of the inspection, the availability of equipment, conditions at the facility, OCM policies, and whether advance notification of an inspection has been given. Documents necessary for the inspection should be prepared in advance of the inspection whenever possible.

The inspector should obtain copies of the inspection forms that are needed for the inspection. Several spare copies of each form should always be carried. FIFRA GLP inspections require:

- FIFRA Notice of Inspection [EPA Form 3540-2]
- FIFRA Receipt for Samples [EPA Form 3540-3].

Forms needed for a TSCA GLP inspection include:

CHAPTER TWO

- TSCA Notice of Inspection [EPA Form 7740-3]
- TSCA Inspection Confidentiality Notice [EPA Form 7740-4]
- TSCA Receipt for Samples and Documents [EPA Form 7740-1]
- TSCA Declaration of Confidential Business Information [EPA Form 7740-2].

In addition, the inspector should be certain to take the following documents and materials on an inspection:

- Copies of FIFRA, TSCA, and the applicable regulations. Inspectors should have copies of FIFRA and TSCA and the applicable regulations with the preamble available upon request. Having such documents available for distribution may help improve the relationship between EPA and the regulated community, which can foster better facility compliance.

- EPA outreach materials. Inspectors should provide current, relevant educational, guidance information to facility officials upon request or as deemed appropriate by the inspector.

- Administrative information. When on travel, the inspector should take travel authorizations and telephone numbers of travel and procurement personnel who may need to be contacted.

Additional documents could include:

- Q & A documents and GLP Regulatory Advisories
- LDIAD Standard Operating Procedures
- Any related *Federal Register* notices.

The inspector should also obtain enough bound field notebooks for each team member, although often the auditors will bring their own notebooks for the inspection. The inspector should assure that loose-leaf or spiral notebooks, or pads of paper are not used. (This prevents anyone from tampering with the notebook, and removes any doubt about the contents of the notebook should it be submitted as evidence in court.) EPA policy is to use only bound notebooks on inspections. However, some inspectors may use a checklist as part of the inspection documentation. If a checklist is used on a GLP inspection, the inspector must (1) reference the checklist with a number or alphanumeric identifier (unique to that inspection) in the bound field notebook and on the checklist and (2) record the date, name of laboratory, and inspector's initials on each page of the checklist.

Appended to this manual are two checklists developed for use on a GLP compliance inspection (Appendix D and E). These checklists are provided as tools for the inspector and their use is entirely optional. Inspectors using these, or any checklist, are reminded that a checklist is only a guide, and should in no way limit the scope of any inspection or investigation.

On occasion, more than one inspection will be conducted at a facility during the same time period (e.g., separate FIFRA and TSCA GLP inspections or inspections at two or more testing facility organizational sub-units within the facility, such as toxicology and chemistry). In this event, separate sets of documents may be required for each inspection, and separate notebooks may be required for each inspection. The inspector should bear this in mind when assembling the inspection supplies.

2.7.2 Inspection Equipment

The types of equipment that an inspector takes to an inspection site will vary depending upon the nature and extent of the inspection and the type of testing facility to be inspected. The inspector should use her/his best judgment, based on training and inspection experience and on knowledge gained in preparing for the inspection, in determining what equipment is necessary for a particular inspection. The equipment should be well-maintained and in good condition at the time of the inspection. Therefore, prior to each inspection, the inspector should check the equipment to make sure that it is in good working condition.

Since each inspection is unique, no single list of equipment or forms can be devised that will fit every inspection situation. The types of equipment most likely to be needed during a GLP inspection are summarized in Table 2-1.

2.8 Confidential Business Information Considerations

2.8.1 TSCA

Section 14 of TSCA and EPA regulations (40 CFR Part 2) protect CBI from disclosure. CBI includes trade secrets (including process, formulation, or production data), the uncontrolled disclosure of which could cause damage to a facility's competitive position. In general, disclosure of CBI is prohibited; however, there are certain specific and limited exceptions (see 40 CFR Part 2). EPA's Office of General Counsel is responsible for making the final administrative determination as to a CBI claim.

An inspector must notify facility representatives of their right to claim data at the facility as CBI. Because the inspector may require access to CBI before (i.e., while preparing for an inspection), during, and after an inspection, the inspector must be knowledgeable of EPA procedures governing access to, handling of, and disclosure of CBI. The inspector and others who may use the information must have TSCA CBI access authorization, since only authorized individuals may have access to CBI. An inspector may need access to CBI data that a subject facility submitted to EPA or provides during the inspection, as well as information that was collected during a prior inspection.

A CBI-cleared inspector can obtain access to CBI documents at EPA by requesting the information from an appropriate Document Control Officer (DCO). The DCO is responsible for the following:

TABLE 2-1. TYPES OF EQUIPMENT

GENERAL	
• Camera • Pocket calculator • Clipboard • Locking briefcase • Stamp pad • Plastic covers • Disposable towels or rags • Portable typewriter • Bound notebooks • Flashlight and batteries • Pens • Blank pad • Attendance sheets for opening and closing conference • Relevant guidelines (FIFRA) or test rule (TSCA) • Pre-addressed envelopes (e.g., to Document Control Officer)	• Film and flash equipment • Tape measure • Waterproof pens, pencils, and markers • "Confidential Business Information" stamp • Plain envelopes • Polyethylene bags • Portable copying machine • Pocket knife • Computer • Sampling material • Pencil • Highlighters, multicolored • Bound notebooks • Post Its • Study reports • Inspection history • Regulations

SAFETY[1]	GLP PROGRAM MATERIALS
• Safety glasses or goggles • Face shield • Ear plugs • Rubber-soled, metal-toed, nonskid shoes • Liquid-proof gloves (disposable, if possible) • Coveralls, long-sleeved • Hard hat • Plastic shoe covers, disposable • Respirators and cartridges • Self-contained breathing apparatus	• Notification letter • Inspector credentials • Letter credentials for noncredentialed team members • Investigation request • Notice of Inspection, FIFRA or TSCA • Receipt for Samples, FIFRA or TSCA • CBI forms, TSCA • SOPs • Directions/maps • Checklist(s)

[1]In accordance with EPA policy, an inspector may not do field work without first completing an approved health and safety training program. Personnel who use respiratory protection equipment must also complete specialized training, which includes training on protective equipment selection criteria. Program-specific safety training has been developed for field personnel facing particular risks.

- Verifying that the requesting inspector is on the authorized CBI access list

- Providing a copy of the CBI classified document to the inspector

- Determining whether the inspector has secure storage approved by EPA if the inspector is to keep the CBI for more than a day.

It is very possible that most data obtained from EPA Headquarters is CBI. To facilitate the transfer of CBI to the Regions or National Enforcement Investigations Center (NEIC) for use in conducting TSCA GLP inspections, EPA established a policy authorizing telephone discussion of CBI information obtained by EPA. Basically, communication by telephone can include CBI when the contact is from Region to Region, Region to specified Headquarters offices, Headquarters to Region, Headquarters or Region to subject facility, or Headquarters to NEIC.

Whether or not it is anticipated that CBI documents will be collected during a TSCA GLP inspection, the inspector must provide a TSCA Inspection Confidentiality Notice to the responsible facility official at the beginning of the inspection. This form is used to inform facility officials of their right to claim part of the inspection data as CBI. The inspector should be familiar with the procedures for asserting a CBI claim and the four criteria which the claimed information must meet. These four criteria are discussed in detail on the TSCA Inspection Confidentiality Notice.

If documents are collected for which facility officials exercise their right to claim confidentiality, the inspector must list all such documents on the TSCA Declaration of Confidential Business Information. The inspector must take custody of all CBI documents before leaving the facility and must maintain them in his/her custody, using all proper procedures and safeguards, until they can be received by a DCO. The DCO will then properly transfer such documents as needed to the auditors or the inspector.

2.8.2 FIFRA

If a facility claims that certain business information is confidential under FIFRA, the inspectors must follow procedures for handling FIFRA sensitive information (e.g., CBI, trade secrets) found in the FIFRA *Information Security Manual* (July 1988).

In general, during an inspection, representatives of a facility inspected under FIFRA must be informed of their right to claim any information as CBI. In the event that documents are collected for which facility officials exercise their right to claim confidentiality, the inspector must identify all such documents (and samples) on the FIFRA Receipt for Samples as confidential. The inspector must take custody of all CBI documents before leaving the facility and must maintain them in his/her custody, using all proper procedures and safeguards, until they can be received by a DCO. The DCO will then properly transfer such documents as needed to the auditors or the inspector.

3.0 INSPECTION PROCEDURES

3.1 INTRODUCTION

This chapter discusses the required procedures for entering a testing facility to conduct a GLP inspection under FIFRA and TSCA. This chapter also presents the statutory requirements, and OCM policies applicable to GLP inspections. Guidance is provided to the inspector on the preliminary aspects of a GLP inspection, ranging from a discussion of the importance of obtaining testing facility owner/operator consent for the inspection to a listing of procedures for conducting the opening and closing conferences. Procedures for determining compliance with various sections of the GLP regulations are covered in succeeding chapters and reference SOPs.

3.2 INSPECTION AUTHORITY

An inspection may be conducted only after the inspector has presented the following items to the owner, operator, or agent in charge:

- Appropriate credentials
- Written Notice of Inspection.

Inspections must be conducted in the following manner:

- Initiated and completed in a timely manner
- Conducted at reasonable times, within reasonable limits, and in a reasonable manner.

3.2.1 Credentials

Section 8 of FIFRA and Section 11 of TSCA require the EPA Administrator or any duly designated representative of the Administrator who conducts an inspection of a testing facility to present the owner, operator, or agent in charge with appropriate credentials and with a written Notice of Inspection (discussed in Section 3.2.2). Credentials are issued by the EPA Administrator or her/his designee and the inspection may be made *only* upon presentation of such credentials. The credentials are identifying papers that state the holder of the papers (i.e., the inspector) is authorized to conduct official investigations and inspections pursuant to all laws (including GLP inspections) that EPA administers. Inspectors must present credentials whether or not testing facility officials request identification. Once testing facility officials have viewed the credentials, they may wish to telephone EPA Headquarters or the Regional Office to verify the inspector's identification, which is permissible. However, credentials may not be photocopied. Credentials should also be readily available so that they can be presented to other facility representatives during the course of the inspection. Lastly, the inspector should make a note in her/his field log that credentials were presented.

EPA employees who do not have EPA inspector credentials, but who accompany the inspector and have a role on the inspection team, such as a data auditor, should be prepared to present their EPA badges or another form of identification (if not an EPA employee) plus a letter of authorization specific to the inspection being conducted. The issuance of such letters for sensitive inspections is merely an additional precaution and does not diminish the authority of an inspector to utilize any appropriate EPA employee as part of an inspection team. The inspector should arrange for and have available letters of introduction and authorization for any members of the inspection team who are not EPA employees.

3.2.2 Written Notice of Inspection

As soon as the inspector has identified herself/himself and presented the required credentials, s/he must present to the owner, operator, or agent in charge of the testing facility with a written notice of "the premises or conveyances to be inspected." This Notice of Inspection (see Figure 3-1 for the TSCA Notice of Inspection and Figure 3-2 for the FIFRA Notice of Inspection) informs the owner, operator, or agent in charge of the reason for an inspection under Section 8 of FIFRA or Section 11 of TSCA and contains the inspector's address and signature. Although the *time* of the intended inspection is not required, the scheduled time (and date) of the inspection should be included in the Notice of Inspection to establish that the inspection was requested at a *reasonable time*, as required by Section 8 of FIFRA and Section 11 of TSCA.

The inspector is required to present the Notice of Inspection after arrival at the testing facility. A separate inspection notice must be presented for each inspection, but a separate notice is *not* required for each *entry* made during the period covered by the inspection (e.g., for each separate day of a multi-day inspection). The inspector should make a note in the inspection notebook that the Notice of Inspection was presented and should include a copy in the inspection report.

3.3 PRELIMINARY STEPS FOR CONDUCTING A GLP INSPECTION

The inspector should use the following procedures to assure compliance with statutory requirements and OCM policy covering GLP inspections:

- To comply with the statutory requirement that the inspection of a testing facility subject to GLP regulations be conducted at a reasonable time, the inspector should inspect the testing facility only during normal working hours, unless mitigating circumstances required that an inspection be conducted during nonbusiness hours. For example, it may be preferable to enter a testing facility that operates on a shift basis during one of the off-hour shifts. Similarly, it may be preferable to start the inspection of a field site early in the morning.

- The inspector should arrive at the testing facility at the time designated (if it has been included) in the Letter of Advance Notification. GLP regulations do not require that the written Notice of Inspection include the time of the inspection, but the inspector may want to include the time on the Notice of Inspection as documentation.

INSPECTION PROCEDURES

FIGURE 3-1.
TSCA NOTICE OF INSPECTION

⊕EPA
US ENVIRONMENTAL PROTECTION AGENCY
WASHINGTON, DC 20460
TOXIC SUBSTANCES CONTROL ACT
NOTICE OF INSPECTION

Form Approved
O.M.B. No. 2070-0007
Approval expires 8-31-85

1. INVESTIGATION IDENTIFICATION			2. TIME	3. FIRM NAME
DATE	INSPECTOR NO.	DAILY SEQ. NO.		

4. INSPECTOR ADDRESS	5. FIRM ADDRESS

REASON FOR INSPECTION

Under the authority of Section 11 of the Toxic Substances Control Act:

☐ For the purpose of inspecting (including taking samples, photographs, statements, and other inspection activities) an establishment, facility, or other premises in which chemical substances or mixtures or articles containing same are manufactured, processed or stored, or held before or after their distribution in commerce (including records, files, papers, processes, controls, and facilities) and any conveyance being used to transport chemical substances, mixtures, or articles containing same in connection with their distribution in commerce (including records, files, papers, processes, controls, and facilities) bearing on whether the requirements of the Act applicable to the chemical substances, mixtures, or articles within or associated with such premises or conveyance have been complied with.

☐ In addition, this inspection extends to *(Check appropriate blocks)*:

☐ A. Financial data ☐ D. Personnel data

☐ B. Sales data ☐ E. Research data

☐ C. Pricing data

The nature and extent of inspection of such data specified in A through E above is as follows:

INSPECTOR SIGNATURE	RECIPIENT SIGNATURE	
NAME	NAME	
TITLE	DATE SIGNED	TITLE	DATE SIGNED

EPA Form 7740-3 (12-82)   * U.S. GOVERNMENT PRINTING OFFICE: 1983-661-893   **INSPECTION FILE**

FIGURE 3-2.
FIFRA NOTICE OF INSPECTION

U.S. ENVIRONMENTAL PROTECTION AGENCY	ADDRESS *(EPA Regional Office)*		
NOTICE OF INSPECTION	DATE	HOUR	A.M. P.M.

NAME OF INDIVIDUAL	TITLE
FIRM NAME	FIRM ADDRESS *(Number, Street, City, State and Zip Code)*
SIGNATURE OF EPA EMPLOYEE	TITLE

REASON FOR INSPECTION

☐ FOR THE PURPOSE OF INSPECTING AND OBTAINING SAMPLES OF ANY PESTICIDES OR DEVICES PACKAGED, LABELED, AND RELEASED FOR SHIPMENT, AND SAMPLES OF ANY CONTAINERS OR LABELING FOR SUCH PESTICIDES OR DEVICES, IN PLACES WHERE PESTICIDES OR DEVICES ARE HELD FOR DISTRIBUTION OR SALE *(Sec. 9(a) and 12(a)(2)(B))*.

☐ FOR THE PURPOSE OF INSPECTING AND OBTAINING COPIES OF THOSE RECORDS SPECIFIED IN SECTION 8 AND 40 CFR PART 169. *(Sec. 8 and 12(a)(2)(B))*.

VIOLATION SUSPECTED:

Section 8, 9(a) and 12(a)(2)(B) of the Federal Insecticide, Fungicide, and Rodenticide Act, as amended *(7 U.S.C. 136 et seq.)* are quoted on the reverse of this form.

EPA Form 3540-2 (Rev. 3-77) PREVIOUS EDITION MAY BE USED UNTIL SUPPLY IS EXHAUSTED

Original - ESTABLISHMENT COPY
1 - SAMPLE RECORD COPY
2 - REGION COPY
3 - INSPECTOR'S COPY

- The inspector should enter the testing facility through the main entrance. If SSB sent an advance letter (or notice) of inspection to the testing facility and in response the testing facility designated a different entrance, the inspector should use the designated entrance.

- Upon arrival at the testing facility, the inspector should identify him/herself to the guard or receptionist, as necessary. FIFRA and TSCA require only that credentials be presented to the "owner, operator, or agent in charge," but the inspector may present them to others as a courtesy or to expedite entry. Inspector credentials should be kept accessible throughout the inspection. Business cards, EPA identification cards, and EPA badges *do not* constitute credentials but may be used for introductory purposes as appropriate.

- The inspector should ask the guard (or the secretary or receptionist, as appropriate) to see the testing facility owner, operator, or agent in charge.

- The inspector may sign a visitor's log or register if requested to do so. However, the inspector *may not* sign any such sheet or other document if it contains restrictive language limiting the scope of the inspection, the manner in which the inspection may be conducted, or the manner in which information may be used (e.g., a waiver, release of liability, restriction on the use of photographic or other recording devices). Attempts to restrict the inspection are discussed in detail in Section 3.4.3.

- Once the owner, operator, or agent in charge has been located, the inspector must present the testing facility official with proper inspector credentials and the FIFRA and/or TSCA Notice of Inspection. The inspector must assure that the person to whom the credentials are presented is the authorized facility representative. If the person greeting the inspection team is not the same individual who communicated with the inspector during the pre-inspection preparation (if such communication occurred), the inspector should verbally determine that this person is authorized to grant entry and allow the inspection to proceed. The verbal acknowledgement of responsibility on the part of the facility representative should be noted in the inspector's field notebook.

- EPA policy requires the GLP inspector to obtain consent from the testing facility owner, operator, or agent in charge prior to entering a testing facility to conduct an inspection. To document consent in the case of a FIFRA inspection, the inspector should record the name and title of the person giving consent and the date and time consent was given in the field notebook. In the case of a TSCA inspection, the facility representative must sign the TSCA Notice of Inspection form. If an inspector is denied entry to a testing facility to conduct a GLP inspection, or if the testing facility owner, operator, or agent in charge attempts to impose conditions upon the entry (i.e., conditional entry), the inspector should follow the actions described in Section 3.4.3.

3.4 Consent To Enter And Inspect

3.4.1 Procedures to Gain Consent

EPA procedure calls for the inspector to obtain consent for the inspection. Therefore, the inspector should not conduct an inspection until s/he has accomplished one of the following tasks:

CHAPTER THREE

- Obtained consent prior to conducting the inspection and/or

- If unable to obtain consent, followed proper procedures governing what should be done when entry to a testing facility is denied or when conditions or restrictions are imposed by testing facility officials upon authorized GLP inspectors (see Sections 3.5 and 3.4.3, respectively).

Consent to enter must be given knowingly and freely. The inspector must not coerce or lie to testing facility officials to induce consent. For example, the inspector must not suggest that failure to permit entry may result in the imposition of civil or criminal penalties, since the entry may then be considered nonconsensual (i.e., because of the "threat" of negative consequences).

Consent must be given by the person in possession (i.e., the owner or operator) of the premises, or by some other person with authority to give consent at the time of the inspection. An owner does not always have possession, and so may not be authorized to give consent; for example, the owner may be renting the property to a tenant who is on the property. If someone in possession of the testing facility cannot be located, the inspector must make a good faith effort to determine who may otherwise consent to the entry (i.e., the agent in charge). The inspector should present her/his credentials to that individual and record the name and title of that person in the field notebook.

The GLP inspector must obtain consent to enter the premises for each noncredentialed person (e.g., study auditor, Senior Environmental Executives (SEEs), observer, or contractor) who intends to accompany the inspector on an inspection. If consent is not given, these persons may not enter the premises to be inspected. However, refusal to allow an inspection team member access to the facility is considered conditional consent. No uncredentialed person may have access to TSCA CBI unless s/he has been authorized for CBI access by the Information Management Division (IMD) of the Office of Pollution Prevention and Toxics (OPPT). IMD circulates monthly a list of persons with TSCA CBI clearance to the Regional and program DCOs.

In most instances, if the inspector follows proper procedures upon arrival at the testing facility (i.e., presentation of credentials and Notice of Inspection), obtaining consent to enter will be simple. However, special situations may arise:

- Owner or operator reluctance to give consent
- Conditional consent
- Withdrawal of consent.

These situations and the inspector's responsibilities in each situation are discussed below.

3.4.2 Reluctance to Give Consent

The receptiveness of testing facility officials toward an inspector may vary from facility to facility. If consent to enter a testing facility is denied outright, the inspector should follow the procedures governing denial of entry (see Section 3.5). However, in some instances, testing facility representatives express reluctance rather than actual refusal to allow an inspector to conduct an inspection. Often this reluctance is due to a misunderstanding by testing facility representatives of EPA and/or testing facility responsibilities under GLP regulations, or to concerns over perceived inconveniences to the testing facility operations as a result of the inspection. Such concerns can often be resolved through diplomacy and discussion but should not be subject to negotiation.

In such instances, an inspector should tactfully ascertain the reasons for testing facility officials' reluctance and try to resolve the problems. The inspector should also explain in detail the purpose of the inspection. However, it is crucial during any discussions that an inspector not agree to any restrictions on the scope of a GLP inspection. In addition, the inspector must avoid issuing threats of any kind or making inflammatory statements.

If entry is still denied, the inspector should withdraw from the testing facility and contact her/his supervisor for further instructions. Under no circumstances should the inspector attempt to gain entry or consent to enter by coercive actions or by making statements that suggest that the testing facility officials could be fined or otherwise "punished" unless entry is allowed.

3.4.3 Conditional Consent

Conditional consent refers to any attempt by testing facility officials to restrict the inspection after entry or to condition entry upon the adherence to restrictions imposed by the officials. The following are some of the more common types of conditions that testing facility officials try to impose on the inspector:

- Waivers, indemnity agreements, or releases
- Photographic or tape recording restrictions
- Requirements concerning health and safety gear or training
- Denial of access to certain areas of the testing facility.

Generally, any attempt by testing facility officials to cause inspectors to deviate from standard inspection procedures should be interpreted as an effort to impose conditions on consent. OCM policy is that inspectors should never agree to restrictions or conditions on the scope or nature of the inspection. The inspector must never sign any document that could compromise EPA's right to conduct an inspection. Any effort by testing facility officials to impose one or more conditions should be considered denial of consent. In such circumstances, the inspector should follow the procedures governing denials of consent (see Section 3.5). However, reasonable requests (such as requesting the inspector to wear a visitor's

CHAPTER THREE

badge, hard hat, or safety glasses) should not generally be considered "conditions," and the inspector should comply with such requests.

Even if the inspector is permitted to enter the testing facility without being asked to comply with conditions or restrictions, the inspector should be constantly aware of any attempt by testing facility representatives to impose such conditions *after* entry. If this should occur, the inspector must regard this attempt as a denial of consent and follow appropriate procedures (continued in Section 3.5).

Some areas in which testing facility officials may attempt to impose restrictions are discussed below.

- <u>Waivers and other restrictive agreements</u>. EPA inspectors have the right and the responsibility to refuse to sign any agreement or waiver that promises that records or other data obtained from the testing facility will not be released to a third party or in documentary form. Any attempt by testing facility personnel to restrict inspection activities by requiring inspectors to sign such restrictive agreements should be viewed as a denial of consent to inspect the testing facility and treated accordingly. (However, inspectors should reiterate to the testing facility owner, operator, or agent in charge of her/his right to claim such data as CBI or a trade secret and of the procedures for making such claims.)

- <u>Restrictions on use of photographic or other recording equipment</u>. Inspectors may document evidence of potential violations at the testing facility by means of tape recordings, photography, or recording by electronic devices with a visual taped readout, or by other methods. Testing facility officials often attempt to restrict the use of any or all of such devices by EPA inspectors. Any attempt by a testing facility representative to restrict the use of such devices is considered a denial of consent and appropriate procedures governing such denials should be followed. Testing facility officials should be advised that photographs and other information and data gathered by recording equipment may be claimed as CBI (see Section 3.7.3).

- <u>Health and safety restrictions</u>. The inspector should ascertain the applicable safety requirements before the inspection, if possible. The inspector should be aware that s/he is subject to the applicable safety requirements of the testing facility. For example, if safety boots and glasses are required to walk through a portion of the testing facility, then the inspector must wear them. However, EPA inspectors cannot be required to participate in the testing facility's safety training program (possibly very time consuming) as a condition of conducting a GLP inspection. If testing facility officials make such a demand, the inspector should refuse and treat the situation as a denial of consent.

- <u>Refusal to allow access to certain areas of the testing facility</u>. If, during the course of the inspection, access is denied or restricted to certain areas of the testing facility, the inspector should make a notation describing such denial or restriction in the field notebook and identify which portion of the inspection could not be completed due to the denied or restricted access. However, despite the access restriction, the inspector should proceed with the remainder of the inspection. After leaving the testing facility, the inspector should contact her/his supervisor to determine the appropriate action(s) to take.

3.4.4 Withdrawal of Consent

Occasionally, testing facility officials may initially consent to an inspection but later withdraw the consent during the inspection. Consent to the inspection may be withdrawn at any time after entry has been made. OCM policy concerning withdrawal of consent is to view it as an outright denial of consent. In such cases, appropriate procedures should be followed (see Section 3.5). All activities and evidence obtained prior to the withdrawal of consent are valid. Therefore, evidence obtained by the inspector before consent was withdrawn is usable in any subsequent enforcement actions and should be retained by the inspector. The inspector should not return any evidence that has been collected before the withdrawal of consent if asked to do so by the testing facility official.

3.4.5 Inspector Judgment

Whenever it appears that testing facility officials, through statements or actions (such as physically blocking entry), are denying or restricting consent to conduct the inspection authorized by FIFRA or TSCA, the inspector must use independent judgment to determine the actual effect the limitation will have on the inspection.

Sometimes inspectors are able to reach an agreement with testing facility representatives. Inspectors should keep in mind that attempts to negotiate with testing facility officials are based on the independent judgment of the inspectors and, as such, involve risks (e.g., having a court declare later that consent was obtained coercively and, therefore, was not consent). If the inspector is in doubt, s/he should consult the appropriate supervisor, or counsel before proceeding.

The inspector always has the option of departing the testing facility premises and advising his/her supervisor. The inspector can choose to discontinue the inspection at any time after the inspection has begun. The following examples illustrate how inspectors could handle situations that might be considered denials of entry:

- <u>Restrictive language in sign-in book or form</u>. The inspector should draw a line through objectionable language before signing, obtain a photocopy, and make a note in her/his field notebook. The inspector must inform the testing facility officials of the modification s/he made and request that a testing facility official initial the modification.

- <u>Photographs</u>. When testing facility officials state they do not want photographs taken of the testing facility, the inspector can proceed with the inspection, raising the issue of photographs again only when a particular photograph is essential to the inspection. If photography is still not permitted, the inspector can follow the procedures for securing and executing a warrant (provided in Section 3.5).

- <u>Safety training</u>. While inspectors are not required to participate in a testing facility's safety training course prior to entry, if the testing facility has a relatively short safety briefing that will not interfere with the inspector's ability to complete the planned inspection in a timely manner, the inspector may want to attend the briefing. Inspectors

CHAPTER THREE

are sometimes able to learn valuable information about the testing facility's layout and operations at such orientations. If safety or facility liability concerns are raised by the facility representative, the inspector should inform the representative that EPA inspectors are required to undergo health and safety training prior to receiving inspector credentials.

The inspector should be sensitive to efforts that may be made during the course of the inspection to limit or interfere with his/her activities. These may range from the subtle, inefficient wasting of the inspector's time to imposing restrictions (discussed previously). If the conduct of the inspection is being compromised, the inspector should regard the situation as withdrawal of consent and proceed as described in Section 3.5.

3.5 DENIAL OF CONSENT TO ENTER AND INSPECT

The first thing that an inspector must keep in mind when testing facility officials deny consent to conduct a GLP inspection or any portion thereof is to refrain from making any inflammatory remarks or statements that could jeopardize subsequent enforcement actions brought against the testing facility. For example, the inspector must not discuss any potential penalties that are authorized under Section 14 of FIFRA or Section 15 of TSCA for refusal to permit entry or inspection. If such potential penalties were discussed and consequently testing facility officials permitted the inspection to be conducted, officials could later claim such statements were a form of coercion. A court could rule at a subsequent proceeding that some or all of the evidence collected during the inspection is inadmissible. As another example, the inspector should not threaten testing facility officials with the fact that s/he will "get a warrant," even though OCM may later try to secure a warrant to conduct the inspection. This is because a court of law, which later may review the search procedures used by OCM, may interpret the statement as coercive, invalidating the search.

In the event that entry is denied, the inspector should write the following information (which will be helpful should OCM decide to obtain a warrant at a later date) in her/his field notebook:

- Testing facility name and address

- Name, title, and telephone number of person who refused entry

- Name, address, and telephone number of the testing facility's attorney (if available)

- Date and time of refusal

- Reason (if given) for testing facility's refusal to allow the inspector to enter and/or to inspect testing facility

- Description of testing facility appearance, including number of buildings and general observations, such as housekeeping practices

- Any reasonable suspicion that refusal stemmed from desire to prevent discovery of regulatory or statutory violations.

After making the above notations in her/his field notebook, the inspector should immediately leave the premises and contact her/his supervisor, who will confer with the appropriate counsel concerning the desirability of obtaining a warrant. The inspector should assure that the testing facility officials have been given (or offered) a copy of the written <u>Notice of Inspection</u> to show that proper procedures were followed.

Generally, after counsel is contacted, s/he will discuss the matter of the inspection by telephone with testing facility officials in order to resolve the issues surrounding the inspection. If the matter can be resolved in this manner, the inspection should proceed without further delay. However, if counsel is unable to solve the problem, the inspector should discuss with her/his supervisor the possibility of obtaining a warrant. If the decision is made to obtain a warrant, the EPA attorney will contact the U.S. Attorney's office for the district where the facility is located. The EPA Attorney generally will arrange for an Assistant U.S. Attorney to meet with the inspector as soon as possible.

3.6 THE OPENING CONFERENCE

3.6.1 Purpose

Once the inspector has presented her/his credentials and the required notice for conducting an inspection, it is time for the opening conference. The inspector should ask whether a conference room or office is available where s/he can conduct the opening conference and review testing facility records and files. The opening conference provides an ideal opportunity for the inspector to strengthen EPA-industry relations. The inspector's role, in addition to that of assessing compliance at testing facilities subject to GLP standards, should be that of public relations liaison with the regulated community. The inspector can serve in this role throughout the inspection, but especially during the opening and closing conferences.

The inspector should be prepared to demonstrate her/his thorough understanding of the major sections of FIFRA and TSCA (i.e., those under which the compliance reviews are conducted) and applicable regulations. Industry representatives will also be familiar with the requirements of the regulations in order to comply with the law. Therefore, it is crucial for inspectors to be well-versed in the regulations promulgated pursuant to FIFRA and TSCA, including any new regulations, as they are published in the *Federal Register*. The inspector should always have copies of relevant regulations on hand. During the course of an inspection, it also may be necessary to refer some facility questions or requests (e.g., for TSCA-related publications) to the TSCA Hotline (202-544-1404) or to refer other matters to appropriate EPA Region or Headquarters staff.

During the opening conference, the inspector should have the following objectives:

CHAPTER THREE

- Confirm that his/her credentials have been presented and accepted.

- Confirm that consent to enter has been given by the facility official.

- Present either the FIFRA <u>Notice of Inspection</u> or the TSCA <u>Notice of Inspection</u> and the TSCA <u>Inspection Confidentiality Notice</u> as required.

- Inform the facility representatives of the nature of the inspection (i.e., routine or for cause). If it is a for cause or priority inspection, evidence gathered may be jeopardized if the facility has not been informed.

- Introduce and properly identify the members of the inspection team.

- Schedule the inspection activities (including the closing conference).

- Request appropriate documents such as master schedule, training records, SOPs, testing facility general information, floor plans, and organizational charts.

- Request records, such as protocols and raw data, that are required for study audits.

- Select a recently completed study to be used as an aid for the compliance review.

- Request a testing facility walk-through for the entire team.

- Update existing information on file for the testing facility at EPA Regions and Headquarters.

- Establish a rapport with testing facility officials.

- Conduct the meeting in a positive and professional manner.

- Answer questions concerning FIFRA, TSCA, and applicable regulations.

- Do not overstep authority to accommodate testing facility officials (e.g., do not give opinions about the acceptability of testing facility practices or whether the testing facility is in compliance with the applicable regulations).

- Establish ground rules, working hours, etc.

- Obtain history, scope of operations, recent management changes, business affiliations, etc.

3.6.2 Procedures

The inspector should begin the opening conference by outlining inspection objectives in general terms to inform testing facility officials of the purpose and scope of the inspection. The inspector should, if necessary, present the testing facility owner (or her/his designated agent) with the TSCA <u>Inspection Confidentiality Notice</u> (Figure 3-3). It should be noted that no equivalent FIFRA form exists. This

INSPECTION PROCEDURES

FIGURE 3-3.
TSCA INSPECTION CONFIDENTIALITY NOTICE

⊕EPA — US ENVIRONMENTAL PROTECTION AGENCY, WASHINGTON, DC 20460
TOXIC SUBSTANCES CONTROL ACT
TSCA INSPECTION CONFIDENTIALITY NOTICE

Form Approved
OMB No 2070-0007
Expires 3-31-88

1. INVESTIGATION IDENTIFICATION			2. FIRM NAME
DATE	INSPECTOR NO.	DAILY SEQ. NO.	

3. INSPECTOR NAME	4. FIRM ADDRESS
5. INSPECTOR ADDRESS	
	6. CHIEF EXECUTIVE OFFICER NAME
	7. TITLE

TO ASSERT A CONFIDENTIAL BUSINESS INFORMATION CLAIM

It is possible that EPA will receive public requests for release of the information obtained during inspection of the facility above. Such requests will be handled by EPA in accordance with provisions of the Freedom of Information Act (FOIA), 5 USC 552; EPA regulations issued thereunder, 40 CFR Part 2; and the Toxic Substances Control Act (TSCA), Section 14. EPA is required to make inspection data available in response to FOIA requests unless the Administrator of the Agency determines that the data contain information entitled to confidential treatment or may be withheld from release under other exceptions of FOIA.

Any or all the information collected by EPA during the inspection may be claimed confidential if it relates to trade secrets or commercial or financial matters that you consider to be confidential business information. If you assert a CBI claim, EPA will disclose the information only to the extent, and by means of the procedures set forth in the regulations (cited above) governing EPA's treatment of confidential business information. Among other things, the regulations require that EPA notify you in advance of publicly disclosing any information you have claimed as confidential business information.

A confidential business information (CBI) claim may be asserted at any time. You may assert a CBI claim prior to, during, or after the information is collected. The declaration form was developed by the Agency to assist you in asserting a CBI claim. If it is more convenient for you to assert a CBI claim on your own stationery or by marking the individual documents or samples "TSCA confidential business information," it is not necessary for you to use this form. The inspector will be glad to answer any questions you may have regarding the Agency's CBI procedures.

While you may claim any collected information or sample as confidential business information, such claims are unlikely to be upheld if they are challenged unless the information meets the following criteria:

1. Your company has taken measures to protect the confidentiality of the information, and it intends to continue to take such measures.

2. The information is not, and has not been, reasonably obtainable without your company's consent by other persons (other than governmental bodies) by use of legitimate means (other than discovery based on showing of special need in a judicial or quasi-judicial proceeding).

3. The information is not publicly available elsewhere.

4. Disclosure of the information would cause substantial harm to your company's competitive position.

At the completion of the inspection, you will be given a receipt for all documents, samples, and other materials collected. At that time, you may make claims that some or all of the information is confidential business information.

If you are not authorized by your company to assert a CBI claim, this notice will be sent by certified mail, along with the receipt for documents, samples, and other materials to the Chief Executive Officer of your firm within 2 days of this date. The Chief Executive Officer must return a statement specifying any information which should receive confidential treatment.

The statement from the Chief Executive Officer should be addressed to:

and mailed by registered, return-receipt requested mail within 7 calendar days of receipt of this Notice. Claims may be made any time after the inspection, but inspection data will not be entered into the special security system for TSCA confidential business information until an official confidentiality claim is made. The data will be handled under the agency's routine security system unless and until a claim is made.

TO BE COMPLETED BY FACILITY OFFICIAL RECEIVING THIS NOTICE:	If there is no one on the premises of the facility who is authorized to make business confidentiality claims for the firm, a copy of this Notice and other inspection materials will be sent to the company's chief executive officer. If there is another company official who should also receive this information, please designate below.
I have received and read the notice	
SIGNATURE	NAME
NAME	TITLE
TITLE / DATE SIGNED	ADDRESS

EPA Form 7740-4 (12-82)   INSPECTION FILE

3-13

September 1993

CHAPTER THREE

notice informs the testing facility representative of the right to claim any information (e.g., documents, records, physical samples, or other material) collected from the testing facility during the inspection as CBI. In a TSCA GLP inspection, testing facility officials should also be briefed on EPA's TSCA CBI procedures (see Section 2.8.1).

The inspector should then describe the inspection plan (see Section 2.3) to testing facility officials. By describing the inspection plan, the inspector fosters an atmosphere of cooperation between EPA and the testing facility, which is important for ensuring an understanding by both parties of GLP compliance issues. By keeping the discussion of the inspection plan general, the inspector can avoid providing advance warnings to testing facility officials.

During the opening conference, the inspector should review with the facility management the anticipated schedule of the inspection, particularly if this is a first-time inspection. A typical inspection agenda includes:

- Opening conference
- Facility walk-through
- Review of general facility literature and information
- Review of records, including SOPs and training records
- Compliance review using an on-going study as model
- Evaluation of archives
- Evaluation of the Quality Assurance Unit
- Facilities, equipment and records evaluation
- Evaluation of test, control, and reference substance handling and records
- Closing conference, including issuing receipt for samples.

The facility personnel should also be made aware of what the auditor's activities and needs will be while the inspector is conducting the compliance review.

The most important objective of the opening conference is to obtain as much information concerning the particular testing facility's operations and practices as possible. In addition, the inspector should use the opening conference to question testing facility officials about such practices, operations, and any other data that may not have been included in EPA records or that require clarification. The inspector should refer to the testing facility's organizational chart to learn who is in charge of what operations at the testing facility and who to contact for additional information.

To save time, the inspector should resolve logistical issues at the opening conference in preparation for the actual inspection. The following factors should be considered:

- Having a testing facility official accompany inspector. Before the inspection is conducted, make arrangements for a testing facility representative to accompany the inspector on the inspection. The representative should be able to describe testing facility layout and operations and to indicate what data, records, etc., should be claimed as CBI. (Testing facility claims of confidentiality may be made only by an individual with the authority to make such claims, as discussed in Section 3.7.3.)

- Testing facility walk-through. At the conclusion of the opening conference, the inspection team will typically tour or walk through the testing facility. This can serve as part of the inspection or compliance review and it can also serve to familiarize the other team members with the testing facility to make the audits more efficient. Since the inspector will normally have additional time to inspect the testing facility in greater detail, care should be taken to make sure that the testing facility staff does not use an inordinate amount of the team's time on this walk-through.

- Schedule of inspection. Arrange for a schedule of necessary meetings to be developed, based on the inspection plan and the inspector's understanding of the responsibilities of various testing facility officials. This schedule will allow individuals enough time to prepare for discussions with the inspectors. Set a specific time and place for the closing conference (see Section 3.7). This conference will provide a final opportunity to gather information from testing facility officials, to answer questions, and to complete administrative duties. The closing conference will also provide a forum for summarizing the inspection.

- Master schedule; selection of an ongoing or recently completed study. The inspector should select an ongoing study from the master schedule to use as a means of carrying out the GLP compliance review. The inspector should select an ongoing study covered by the regulations in which the inspection is being carried out. For example, if it is a FIFRA inspection, a study that will be submitted under FIFRA should be selected.

- Facility documents. As part of the opening conference, the inspector should indicate those facility documents s/he will need to review during the course of the inspection, so that the facility personnel can have time to gather the materials. These typically include SOPs, staff training records including resumes and curriculum vitae, quality assurance records, etc. The inspector should also request (particularly if this is a first-time inspection of a new facility or if there have been significant changes to the facility since the last inspection) copies of floor plans, organizational charts, brochures, etc., for use in preparation of a complete report.

3.7 THE CLOSING CONFERENCE

3.7.1 Purpose

The purpose of the closing conference is to allow the inspector and testing facility representatives to resolve final administrative matters concerning the inspection. Ideally, the closing conference should be scheduled for the morning after the conclusion of the GLP inspection. The closing conference should accomplish the following:

- Summarize the inspection proceedings.

- Review all significant inspection findings for facility personnel, clarify any issues, and answer any questions related to the findings.

- Complete and submit to testing facility representatives certain forms related to the inspection (see Section 3.7.3).

- Obtain from testing facility officials any outstanding records or other data needed by the inspector.

- Issue a TSCA <u>Receipt for Samples and Documents</u> and/or a FIFRA <u>Receipt for Samples</u> and allow facility officials to make any confidentiality claims for documents collected during the inspection.

- Make specific arrangements for any additional documents or information to be provided to the inspector.

The inspector should be certain that all appropriate facility personnel will be present. At the very least, this should include management and the study directors. Often, however, the facility will also wish to have others attend, such as technical personnel, attorneys, and corporate or sponsor representatives. It is up to the inspector what, if any, limits should be placed on attendance at the closing conference.

3.7.2 Pre-Closing Conference with Inspection Team

Prior to the closing conference, the inspector should meet with inspection team members to review and discuss inspection findings. It is imperative that the inspector be aware of any problems found by the auditors, and s/he may need to provide guidance to the auditors, especially in determining the significance of negative GLP findings relating to the audited studies. The auditors must also make sure that the inspector understands the significance of any technical or data deficiencies in the studies. No "surprises" should arise during the actual closing conference.

The inspector should determine from each auditor if there are any unresolved issues, if any followup information or records must be requested from the testing facility, and if each auditor has obtained copies of all necessary records or data that were requested and are needed to document any negative findings. The inspector should also establish the order in which findings will be presented. Normally, each auditor will present his/her own findings. However, there may be occasions when an auditor is unwilling to speak (e.g., s/he may not be fluent in English) and may prefer to have someone else present the audit findings. The inspector should make these arrangements beforehand.

3.7.3 Procedures

The inspector serves as the moderator for the closing conference. The same objectives that govern the manner in which an inspector should conduct the opening conference (see Section 3.6) apply to the closing conference as well. The inspector should make a record of the facility personnel attending the closing conference, particularly if problems were encountered during the inspection or there are significant negative findings. The inspector and appropriate team members should summarize the inspection findings in an objective and factual manner. During the conference, it is critical that the team refrain from drawing conclusions regarding any potential violations that may have been discovered during the inspection. Enforcement policy and issues should not be discussed. Any questions related to these areas should be referred to Chief, SSB.

The inspector should clarify any final questions and provide the testing facility officials with the opportunity to ask any final questions. If the inspector wants to obtain additional documents not available at the time of the closing conference, the inspector should request these documents and agree upon a reasonable date by which the testing facility officials should submit the documents. However, the inspector should make every effort to obtain all necessary records during the inspection, thus keeping the need for followup information to a minimum. The inspector should request that any documents that the company may decide to submit to EPA Headquarters or correspondence with EPA resulting from the inspection be copied and sent to her/him.

Areas of disagreement between the inspection team and facility personnel may arise during the closing conference. If these cannot be resolved to the satisfaction of all parties, the facility may be given the opportunity to respond to specific inspection findings in writing. The facility is under no obligation to provide written responses and the inspector is not under any obligation to include such a response in the final report. However, misunderstandings often may be resolved in this manner, and a written explanation of an inspection finding may be useful to include in the final report.

A separate issue is that of recording the closing conference. Commonly, the facility representative will request that the proceedings be tape-recorded; less frequently, s/he may request videotaping or verbatim stenography (as by a court reporter). As a general policy, the closing conference *should not* be recorded. However, the inspector has the discretion to make an exception to this policy under special circumstances. If such is the case, the inspector must stipulate that a copy of the recording or transcript be provided to him/her as part of the inspection documentation.

During the closing conference, the inspector must also require the testing facility officials to complete two forms related to the inspection process. These forms are the TSCA <u>Declaration of Confidential Business Information</u> (Figure 3-4), if necessary, and the FIFRA <u>Receipt for Samples</u> and/or the TSCA <u>Receipt for Samples and Documents</u> (Figures 3-5 and 3-6).

FIGURE 3-4.
TSCA DECLARATION OF CONFIDENTIAL BUSINESS INFORMATION

US ENVIRONMENTAL PROTECTION AGENCY
WASHINGTON, DC 20460

TOXIC SUBSTANCES CONTROL ACT

DECLARATION OF CONFIDENTIAL BUSINESS INFORMATION

Form Approved
OMB No 2070-0007
Expires 3 31 88

1. INVESTIGATION IDENTIFICATION			2. FIRM NAME
DATE	INSPECTOR NO.	DAILY SEQ. NO.	

3. INSPECTOR ADDRESS	4. FIRM ADDRESS

INFORMATION DESIGNATED AS CONFIDENTIAL BUSINESS INFORMATION

NO.	DESCRIPTION

ACKNOWLEDGEMENT BY CLAIMANT

The undersigned acknowledges that the information described above is designated as Confidential Business Information under Section 14(c) of the Toxic Substances Control Act. The undersigned further acknowledges that he/she is authorized to make such claims for his/her firm.

The undersigned understands that challenges to confidentiality claims may be made, and that claims are not likely to be upheld unless the information meets the following guidelines: (1) The company has taken measures to protect the confidentiality of the information and it intends to continue to take such measures; (2) The information is not, and has not been reasonably attainable without the company's consent by other persons (other than governmental bodies) by use of legitimate means (other than discovery based on a showing of special need in a judicial or quasi-judicial proceeding); (3) The information is not publicly available elsewhere; and (4) Disclosure of the information would cause substantial harm to the company's competitive position.

INSPECTOR SIGNATURE		CLAIMANT SIGNATURE	
NAME		NAME	
TITLE	DATE SIGNED	TITLE	DATE SIGNED

EPA Form 7740-2 (12-82)

INSPECTION FILE

FIGURE 3-5.
FIFRA RECEIPT FOR SAMPLES

FIGURE 3-6.
TSCA RECEIPT FOR SAMPLES AND DOCUMENTS

US ENVIRONMENTAL PROTECTION AGENCY WASHINGTON, DC 20460		
⊕EPA	TOXIC SUBSTANCES CONTROL ACT RECEIPT FOR SAMPLES AND DOCUMENTS	Form Approved. OMB No. 2070-0007 Approval expires 3-31-88

1. INVESTIGATION IDENTIFICATION			2. FIRM NAME
DATE	INSPECTOR NO.	DAILY SEQ. NO.	

3. INSPECTOR ADDRESS	4. FIRM ADDRESS

The documents and samples of chemical substances and/or mixtures described below were collected in connection with the administration and enforcement of the Toxic Substances Control Act.

RECEIPT OF THE DOCUMENT(S) AND/OR SAMPLE(S) DESCRIBED IS HEREBY ACKNOWLEDGED:

NO.	DESCRIPTION

OPTIONAL:

DUPLICATE OR SPLIT SAMPLES: REQUESTED AND PROVIDED ☐   NOT REQUESTED ☐

INSPECTOR SIGNATURE	RECIPIENT SIGNATURE
NAME	NAME
TITLE	DATE SIGNED	TITLE	DATE SIGNED

EPA Form 7740-1 (12-82)   INSPECTOR'S FILE

The copy of the FIFRA <u>Receipt for Samples</u> or the TSCA <u>Receipt for Samples and Documents</u>, whichever is applicable, lists and describes each document and sample taken by the inspector from the testing facility. This receipt must be signed and dated by a testing facility official and the inspector. The receipt should identify the following:

- A description of all physical samples taken (if any)

- A description of all records, photographs, or other property taken (particularly crucial when inspecting with a warrant)

- A brief description of information claimed as TSCA CBI (which should be listed on the TSCA <u>Declaration of Confidential Business Information</u> form).

The purpose of this detailed receipt is to document that testing facility officials knew exactly what documents and samples were taken and to allow for full review by testing facility officials so that confidentiality claims can be made.

Many times the facility will routinely consider all documents, both FIFRA and TSCA, to be confidential and will stamp them in some way to indicate this. The inspector should not suggest or advise that any document or other item be claimed as confidential. The decision must be left entirely to the testing facility official. CBI claims should be considered for review and challenge in a manner consistent with the requirements of 40 CFR §2.203 and are the responsibility of the Office of General Counsel. IMD personnel can provide assistance to the inspector in reviewing TSCA CBI claims to determine if they are valid.

The TSCA <u>Declaration of Confidential Business Information</u> form must include a list of all documents, photographs, or other data claimed by an authorized testing facility representative as TSCA CBI. The inspector should keep in mind that some data may have been declared CBI during the inspection. These items should be confirmed with testing facility officials and included on the CBI declaration form. Both the inspector and the claimant (i.e., the testing facility official) must sign and date the completed document.

All documents for which a CBI claim has been made must subsequently be handled according to the custody requirements of the CBI regulations. Therefore, CBI documents should be collected only when necessary to document a potential violation, and the inspector must have determined beforehand that s/he or the auditor has the necessary facilities and procedures in place to meet the CBI requirements.

If there is no onsite testing facility official authorized to make CBI claims at the time of the closing conference, the inspector should:

- Make a copy of the completed and signed FIFRA <u>Receipt for Samples</u> or TSCA <u>Receipt for Samples and Documents</u>.

CHAPTER THREE

- Make a copy of the completed and signed TSCA <u>Inspection Confidentiality Notice</u>.

- Complete the top portion only (i.e., investigation identification information, firm name, inspector address, firm address) of the <u>Declaration of Confidential Business Information</u> form.

- Mail the forms *certified, return receipt requested* to the Chief Executive Officer (CEO) of the testing facility within 2 days of the inspection. (The CEO will have 7 calendar days to make CBI claims on the declaration form.)

The inspector need not take any measures during the 7-day period mentioned in the last item above beyond following the routine security procedures normal for inspection data collected from the testing facility. However, immediately upon notification of the DCO by the appropriate testing facility officials (i.e., those authorized to claim CBI) that data are being claimed as CBI, EPA will commence TSCA CBI procedures with respect to such data. If a CBI claim is made, it is the responsibility of the DCO to notify all parties (i.e., inspector, testing facility, case preparation staff, and any others who may be handling the information) of the fact that the material is CBI and to log the material as required.

3.8 EVIDENCE COLLECTION AND ACCOUNTABILITY

The inspector or auditor will document in the inspection report for potential use as evidence all GLP deviations and deficiencies, and all data discrepancies and gaps. The evidence will be used to support any enforcement or regulatory action EPA takes. This evidence may consist of the following:

- Field notebook entries
- Photocopies of raw data, records, reports, or correspondence
- Personal statements
- Photographs
- Sample or specimen analysis or evaluation.

There should be no assertion or description of a GLP or data problem in an inspection report without proper supporting evidence. Occasionally, documents may also be collected strictly for specimen purposes or to verify correction of a previously identified deficiency. Copies of data, records, etc., may also be collected to allow further detailed review by the inspector, auditor, or other designated expert at a later date. These documents may or may not eventually be used as evidence. In all cases, collection and subsequent handling of these materials must be done using the procedures described here. Inspectors should review the relevant chapters of the *Pesticides Inspection Manual* and the *Basic Inspectors Manual* for further guidance on evidence collection and accountability.

3.8.1 Field Notebook

The field notebook or the inspector's/auditor's personal notes will constitute the principal means of organization for evidence collection for each GLP inspection and related study audit. The field notebook will also serve as a source for describing the overall inspection, personal statements, records collected or requested, calculations, and other information. Observations of operations, facilities, equipment, personnel, and other topics will also be recorded in the notebook.

Each inspection team member will have his/her own notebook. A separate notebook should be used to record information for a single facility. Notes from multiple study audits may be recorded in a single notebook as long as all audit findings are distinctly identified and reflect audits conducted at a single test facility.

The first page of the notebook should be identified with the inspector's (or auditor's) name, the facility, the date(s), and if available, the EPA inspection number. Any business cards collected during the inspection may be stapled along the edge of the opening or closing page. They should not be paper-clipped or left loose in the notebook.

Since the field notebook represents the single most important reference for the inspection report and for any subsequent legal support that may arise out of an enforcement action, care is to be taken with detail, legibility, thoroughness, and accuracy. All entries are to be adequately identified and dated, and each person interviewed identified by name, title, and association with the work under review. The field notebook is to contain only facts and observations. Under no circumstances should a field notebook contain any personal opinions or prejudgment as to whether or not a violation has occurred.

The field notebook is considered EPA property and, as such, is to be retained in the appropriate evidentiary files along with the final inspection report and other related documents.

3.8.2 Copies of Data, Records, Reports, and Correspondence

Photocopies of raw data, records, reports, and correspondence will serve as the primary source of evidence when documenting GLP/study audit problems. All documents necessary to prove a violation should have been collected by the inspector. For example, in a FIFRA GLP compliance inspection, the inspector should be sure to obtain the compliance statement of each study. In addition, if an inspection is focusing on an ongoing study, the inspector must collect evidence that the facility knew the study was going to be submitted to EPA.

Most larger laboratories and testing sites have photocopying facilities and will reproduce a reasonable number of copies at no cost. For other facilities, such as smaller laboratories or remote field sites, the inspector should ascertain, prior to the inspection, whether photocopying capability will be available onsite. If not, arrangements must be made to:

CHAPTER THREE

- Have the documents reproduced at a commercial copying facility during the inspection and recover the cost via travel voucher or petty cash, or

- Arrange to transfer the necessary original documents to the EPA office for reproduction.

In either case, an appropriate FIFRA or TSCA receipt (see Chapter 2.8) should be issued if the documents are removed from the inspected facility, unaccompanied by a facility representative. If documents are copied at the EPA office, the originals should be immediately returned to the facility by hand delivery or by certified mail. In either case, a written acknowledgement of their receipt by an official from the facility should be obtained.

When photocopies of documents are received or made by the inspector, each page must be examined and compared to the original to assure it is an acceptable copy and that the GLP deviation or data discrepancy is clearly evident. If colored inks were present in the original and were not reproduced on the photocopy, the inspector should indicate the colored portion on the copy via arrows or encirclement, and should initial and date the notation(s) in contrasting ink. If the potential violation involves an entry in pencil, a white-out, or other circumstance that is not evident on the photocopy of the document, then the inspector should indicate the deviation with an arrow and/or encirclement and explain it briefly on the document. If possible, these notations should be initialed and dated by both the inspector and a witness. The witness should preferably be another EPA employee or representative; however, it may be a facility official or third party if circumstances warrant. The inspector's and witness's notations should be made with a contrasting colored ink. An arrow stamp can also be used in similar fashion to highlight a particular data or procedural discrepancy on a photocopied document. In addition, each document (or document set) should be listed on the FIFRA Receipt for Samples or TSCA Receipt for Samples and Documents at the close of the inspection (see Chapter 3.7.3).

Related documents should be combined, as appropriate, to form a document set. The various sheets, consisting of copies of raw data for a particular study, may be one document set; a copy of the protocol would be considered a separate document set. The location of the original documents in the facility's archived files should be fully described in the field notebook so that these materials may be readily located at a later date, if necessary.

The reverse side of the first page of each collected document set is to be stamped or identified as follows, with the appropriate information filled in:

Inspection: _____
Doc. No: _____
No. Pages: _____
Date: _____ By: _____

GLP Inspection Manual

The Doc. No. should correspond to the listing on the TSCA or FIFRA receipt. Every page of each document set is then identified on the front in the lower righthand corner with the inspector's initials and the date of collection.

The inspector will collect and maintain all collected and identified documents as a master set of exhibits for each official inspection or investigation. Additional copies may then be distributed to the auditors, either before leaving the site (if multiple copies can be conveniently obtained) or after photocopying upon the inspector's return to the office. In the latter case, the necessary additional copies should be transmitted to the auditor(s) as expeditiously and securely as possible.

Followup information or documents received by the inspector after the onsite phase of the inspection is completed will be handled in the same manner (see Section 6.2).

When a document set (or a portion thereof) is ready for insertion as an exhibit into the final inspection report, each page shall be additionally marked with the exhibit number and the exhibit page number.

3.8.3 Signed Statements

In some circumstances, a signed personal statement from a facility representative may be necessary to fully document an inspection problem, such as in the case of unretained raw data or when potential GLP violations or data problem(s) cannot be fully documented by other means. To the extent possible, signed statements that document any potential violation should be collected.

The statement, which may also be in the form of a letter, should be prepared and signed by a responsible management official or study director who represents the facility or company. Statements may also be obtained from technical and QAU staff, as appropriate. All personal statements should be acknowledged via official FIFRA or TSCA receipt or by letter.

Of particular concern to the EPA inspector or auditor is the case of unretained or missing raw data. Not only does this situation constitute a potential violation of the GLP regulations (40 CFR §160.195 and §762.195), but in the case of data related to registered pesticides, it is also a potential violation of the FIFRA books and records regulations [40 CFR §169.2(k)]. Thus, both sections should be cited in the inspection report when registered pesticide products are involved.

A signed statement from a test facility representative should also be obtained, if possible, when a potential violation is discovered as a result of an interview. The substance of the allegation may be documented in a statement, prepared and signed by the individual, that clearly describes the situation in question. In addition, substantiating evidence in the way of document copies or photographs should also be collected.

3.8.4 Photographs and Observations

In some situations, a photograph or series of photographs is the most effective or only way to document a GLP deviation. Photographs may also be used to enhance the effectiveness or validity of other documentation. Problems associated with facilities, test substance description, equipment, or specimen/reagent labeling, are examples of areas that may lend themselves to photographic documentation.

Some facilities or corporations have policies prohibiting photography or possession of cameras on company property. If photographs are necessary, arrangements can usually be made with facility management to provide a duplicate set of photos or to have facility personnel take side-by-side photographs. Assurance of confidentiality regarding the necessary photographs may also expedite permission. If permission is still refused, this is considered denial of consent (see §3.4.3), and supervisory and/or legal assistance should be obtained by the inspector, particularly if alternative evidence is not available.

Either an instant print (e.g., Polaroid) or 35-millimeter camera may be used; however the latter will provide better quality prints for use as evidence and can be enlarged, if necessary. An instant print camera has the advantage, however, of providing an immediate image to assure that the necessary information was captured. In critical situations, both cameras should be used, with the instant film prints serving as backup if the 35-millimeter prints prove to be unsatisfactory. If a declaration of TSCA CBI is made (or is anticipated), photographs should be taken with an instant print camera, particularly if a developing facility cleared for TSCA CBI is not readily accessible.

All photographs taken as part of an inspection should be fully described in the field notebook with date and time, subject matter, distance, direction, witnesses, exposure data, film type, and other supporting information. Some cameras will automatically provide date and time, as well as sequence number on the negative. When prints are received, whether instant film or 35-millimeter, each should be identified in ink on the back or in the margin with print number, date, inspector's initials, facility, and brief description of subject matter. Computerized label-making systems may also be used.

All photographs should be described on the FIFRA or TSCA receipt, including a notation if duplicate prints are to be (or have been) provided to the facility.

The following references provide the inspector with some introductory technical guidance when taking photographs for use as evidence.

- "Fundamentals of Photography for Government," 1986, U.S. EPA Region 10, Seattle, WA

INSPECTION PROCEDURES

- "Basic Inspector Training Course: Fundamentals of Environmental Compliance Inspections," Chapter 15B, 1989, U.S. EPA, Office of Enforcement, Washington, DC

- "Federal Law Enforcement Training Center: Student Text, Photography ST-39.1/.2," 1977, U.S. Department of the Treasury, Washington, DC.

3.8.5 Physical Sampling (Nondocumentary Samples)

Although the inspector will rarely need to collect nondocumentary samples (i.e., physical samples) as part of a GLP inspection, he/she should be aware of the techniques of sampling, procedures for identification and custody, and the availability of subsequent storage or analysis. Directed investigations are more likely to require sample collection than are routine compliance inspections and audits. In any event, if physical sampling is contemplated, the supervisor should first be consulted regarding issues of authority, analysis, and safety.

Examples of different types of situations in which an inspector may need to collect a physical sample as evidence include:

- Test or control substances. (1) The physical appearance does not agree with the written description in protocol, final report, or raw data; (2) a significant impurity is suspected to have been misanalyzed in the test substance; or (3) the test substance was reported as being technical grade, when it appears to have actually been reference grade material.

- Reference substances. There is reason to question either the stated identity or purity of the reference substance.

- Analytical specimens. (1) Specimens are improperly labeled and a photographic exhibit is not feasible; (2) the actual specimens appear to differ from the matrix described in the protocol, raw data, or the study report; or (3) there are serious concerns about the validity of the reported findings, and analytical confirmation by EPA is feasible and warranted.

- Slides, blocks, or wet tissues. Photographic reproduction or other documentation is not feasible or available and (1) there is a discrepancy between the reported findings of the facility's pathologist/toxicologist and the findings of EPA's auditor; (2) specimens are not labeled according to GLP requirements; or (3) specimens appear to have deteriorated appreciably due to improper archival conditions.

    Since blocks or specimens will represent original materials that cannot be reliably sampled with representativeness preserved, the supervisor should be consulted before sampling. An additional slide can usually be prepared from the original block, if necessary.

- Test system feed, water, soil, or bedding. (1) Unreported physical or chemical contamination that may have affected study integrity is suspected; (2) the feed, soil, water, or bedding does not appear to conform with the protocol, applicable SOP, raw data, or study report, and this problem cannot be documented by other means.

- Test, control, or reference substance mixtures with carrier. (1) There are concerns or suspicion that test or control substance mixtures with carrier are not stable over the period of their use or that they may not have been homogeneous; (2) mixtures are suspected of having been improperly analyzed; or (3) there is suspicion that the wrong or a degraded test, control, or reference substance was used.

Guidance on collecting certain types of chemical samples and chain-of-custody is provided in the TSCA and FIFRA inspection manuals. In all cases, the inspector should consider the following prior to sampling:

- The sample is absolutely required for use as evidence and other forms of documentary samples cannot be used or are not available.

- Means are available to properly and safely collect the sample, preserve it (if necessary), and transport it to the testing facility or other EPA designated facility.

- Proper chain-of-custody can be initiated and maintained.

- The appropriate analyses or other evaluation can be performed in a reliable, timely, and legally defensible manner.

Specific guidance on sampling procedures is provided in SOP GLP-S-02 (in preparation) when collecting physical samples intended to document potential TSCA or FIFRA GLP violations.

3.8.6 Maintenance of Inspection Materials

The security of all materials resulting from an inspection must be maintained by the inspector to assure their integrity for possible use as evidence.

These materials include, but are not limited to:

- Completed inspection forms and notification letters
- Field notebook(s)
- Document copies
- Signed statements
- Photographs
- Physical (nondocumentary) samples/specimens.

Materials must remain secure during two stages: (1) while in the inspector's possession during report preparation and (2) after transfer to longer-term office storage or archiving. The inspector is responsible for maintaining the master file of the original versions of documents or photographs collected during (or in followup to) an inspection. Copies of these materials, not the originals, should be made available to auditors for their use in preparing study audit reports. Physical samples and other evidence are also the inspector's responsibility, particularly with respect to establishing and maintaining custody when further analysis or evaluation is required.

Most evidence collected will be in the form of data or record copies, written statements, field notebooks and photographs — materials that can be secured by the inspector in a locked file cabinet or desk. At a minimum, during inspection report preparation, the collected evidentiary materials should be maintained in a secured office. Only those with a "need to know" should view the materials and then only with knowledge of or in the presence of the inspector. Auditors are to take the same general security precautions with copies of data and records in their possession.

Once the inspection report has been prepared, signed, and transmitted to OCM, the inspection materials not included in the report should be placed in a longer-term evidence file along with a copy of the report, the field notebook(s), and any other related documents. Auditors' notebooks should also be recalled and placed in the appropriate OCM, NEIC, or Regional permanent evidence files, particularly when the auditor is not an EPA employee or not a regular member of the GLP inspection staff.

The permanent evidence file for each office (LDIAD, NEIC, or the Region) should be organized for ease of retrieval either by facility or chronologically (or a combination thereof). An individual should be designated to manage the longer-term storage of evidence files to assure their completeness, retrievability, and integrity.

3.8.7 FIFRA/TSCA Receipts

All evidence collected at a facility during an official inspection is to be listed or described on an appropriate FIFRA or TSCA receipt form before the inspection team leaves the facility. All photocopied data and records, brochures, photographs, and physical samples are to be listed. Related data and records should be grouped together and stapled to form a document set that can be more conveniently listed on the appropriate receipt. Each page or document set should be marked as previously described, and the total number of pages noted on the receipt.

3.8.8 Investigations Involving Alleged Criminal Activity[**]

Participants in criminal investigations may be subject to additional requirements governing such investigations. Because of the severe penalties that may be imposed on the individuals convicted of violating the criminal provisions of the environmental laws or other statutes, there is closer scrutiny of constitutional safeguards to protect the individual's rights. Special Agents of the EPA, Criminal Investigations Division, will provide instructions regarding these safeguards to the project team on all investigations in which they are involved. From the beginning of such an investigation until it is completed, the rights of all individuals must be fully protected.

[**] Taken from "NEIC Policies and Procedures," August 1991, EPA-330/9-78-001-R, U.S. EPA National Enforcement Investigations Center, Denver, CO.

If, during the course of a civil/administrative inspection, aspects of criminal activity become apparent, the inspector should obtain all the evidence documenting the possible violation. The Criminal Investigations Division or the appropriate Special Agent-in-Charge must then be apprised immediately. If a criminal investigation is opened, any files will be maintained separately from the current (or planned) civil investigation. Where applicable, EPA's policy on parallel criminal and civil enforcement proceedings must be followed.

4.0 GLP Compliance Review

4.1 Introduction

The compliance review is that part of the inspection in which the inspection team attempts to determine the extent to which the facility is complying with the principles set down in the GLP regulations at the time of the inspection. The conduct of a compliance review generally requires a walk-through of the facility and an audit of the records and procedures of an ongoing study or recently completed study. Depending upon the situation at the testing facility, the review may be altered, based on the needs of the inspection team, immediate availability of the testing facility or staff to be visited, studies being conducted, or other circumstances.

Major areas of concern include:

- Adequacy of organization (e.g., existence of an independent Quality Assurance Unit QAU)

- Adequacy of staffing, facility, and equipment for the workload

- Completeness and adequacy of SOPs and protocols

- Adequacy of archives

- Credentials of staff to perform work as shown by curriculum vitae, training records, and other documents

- Receipt, storage, and use of test, control, and reference substances

- Receipt, handling, care, and use of test systems

- Records of analysis for contaminants of food, water, soil, and other media.

Inspectors should review the contents of relevant SOPs, such as SOP GLP-C-01, Conducting a Field Site Compliance Inspection, for more detailed instruction on compliance reviews. The inspector should be aware that EPA may from time to time issue new Standard Operating Procedures (SOPs). These SOPs are the primary guidance documents developed by EPA to inform GLP inspectors of current policy and procedures. As such, all current SOPs will take precedence over the contents of this manual. In areas where this manual and newer SOPs differ, the directives of the SOPs will be followed.

4.2 Facility Compliance Review

The inspector should examine the master schedule to select an ongoing study for use as a basis for the compliance review. By tracking the progress of the study, the records being taken, and the compliance of the staff with the SOPs, it is possible to evaluate the facility's current level of compliance with the

CHAPTER FOUR

GLP regulations. In addition, the inspector should examine the layout of the facility and the organizational charts of the facility to get an understanding of the scope of the facility. This will assure that the familiarization walk-through of the facility covers all appropriate areas. During the walk-through, the inspection team members should identify potential areas of concern. They should discuss their concerns and impressions with the lead inspector for followup later in the inspection.

The following information for the ongoing study being used for the compliance review should be recorded during the inspection:

- Applicable statute
- Test substance
- Study title
- Lab project number
- Sponsor
- Sponsor representative
- Study start date
- Study director.

The inspector should select an ongoing FIFRA study for FIFRA inspections and an ongoing TSCA study for TSCA inspections. If that is not possible, the inspector should be sure to give the facility the appropriate notification (i.e., FIFRA or TSCA Notice of Inspection) to cover the selected study. The areas that should be evaluated during the facility compliance review are discussed below.

4.3 ORGANIZATION AND PERSONNEL

4.3.1 Personnel [40 CFR §160.29]

- Education, training, and experience. The inspector should determine whether facility supervisors and staff have the education, training, and experience, or a combination of these, necessary to perform their assigned functions. During the course of the inspection, the inspector should also observe the actions and responses of facility personnel as indications of whether their training and experience are appropriate. The staff's responses to questions concerning the operation of the testing facility or the conduct of the study(ies) observed during the walk-through can provide insight into their competence. The inspector should note any deviations from routine testing facility practices, i.e., spilled samples, careless analyses, dirty facilities, recordkeeping mistakes, or other actions that show the inexperience or inadequacy of the staff. Further, the inspector should note whether records show missing data; this may indicate that the staff is either not attentive to details or lacks the time to record data properly.

- Personnel information. The inspector should examine the facility's summary of the job descriptions, training records, curriculum vitae, and experience for its staff and supervisors. This summary must be maintained and updated by the facility, and should contain enough detail to show whether the qualifications of personnel meet the standards of their jobs.

- <u>Number of personnel</u>. The number of personnel must be sufficient for timely and proper conduct of the study according to the protocol. The inspector should compare the actual number of personnel to the number of personnel deemed necessary by the study protocol. The inspector should note if there are any delays in the processing of test substances, the analyses of specimens, or the completion of reports due to insufficient staffing and/or equipment.

- <u>Personal sanitation and health</u>. Facility personnel must take steps to prevent contamination of test systems and substances. The inspector should observe and inquire about the personal sanitation and health precautions used to prevent contamination of test systems, specimens, or substances or otherwise adversely affecting the quality of the study.

- <u>Appropriate clothing</u>. Facility personnel must wear clothing appropriate for the duties they perform. The inspector should observe whether adequate protective clothing is in use at the time of the inspection. The inspector should also inquire about the facility's procedures for handling protective clothing, including requirements for changing clothing as often as necessary to prevent microbiological, radiological, or chemical contamination of test systems, specimens, and test, control, and reference substances.

- <u>Personnel illnesses</u>. The inspector should determine whether precautions are in place to exclude from the test area any individual who has an illness that may adversely affect the study. Personnel must have been instructed to report such illnesses to their immediate supervisor.

4.3.2 Testing Facility Management [40 CFR §160.31]

The inspector should review the management practices of the facility to determine whether they meet the requirements of the GLP regulations.

- <u>Designation of a study director</u>. Management must designate a study director prior to initiation of the study. The inspector should examine facility records to determine if they show that a study director was designated before work began and whether the study director has the appropriate education, training, and experience.

- <u>Replacement of a study director</u>. Management must replace the study director promptly if necessary. Thus, an ongoing study that is the focus of the inspection must have a study director in place.

- <u>Management assurances</u>. Management must also assure the following:

    - A QAU is in place as required by 40 CFR §160.35, as appropriate.

    - Test, control, and reference substances or mixtures have been appropriately tested for identity, strength, purity, stability, and uniformity, as applicable.

CHAPTER FOUR

- Personnel, resources, facilities, equipment, materials, and methodologies are available as scheduled.

- Personnel clearly understand the functions they are to perform.

- Any deviations from the GLP regulations reported by the QAU are communicated to the study director and corrective actions taken and documented. (While the inspector cannot review the findings of the QAU, the inspector can check to determine if the findings were sent to the study director and that corrective actions were taken. This may be done by reviewing the QAU records that do not include QAU findings, problems or actions recommended and taken. This may also be done by requesting management to certify that inspections are being implemented, performed, documented, and followed up in accordance with §160.35.

If the inspector finds deficiencies in any of these areas, s/he should determine whether management failures contributed significantly to the deficiencies. Deficiencies in facility management may be evidenced by deficiencies in other areas of GLP compliance and may be determined by considering some of the following aspects of the study:

- Whether the facility has sufficient personnel and the organization to complete the study in accordance with GLP standards

- Whether the QAU is independent from the individuals conducting the study

- Whether the QAU reports its findings to the study director and testing facility management

- Whether the QAU maintains a record of the quality assurance inspections that meet GLP requirements.

4.3.3 Study Director [40 CFR §160.33]

The study director has overall responsibility for the technical conduct of the study as well as the interpretation, analysis, documentation, and reporting of results. The study director represents the single point of study control. As such, the inspector must assure that the following items meet the GLP requirements:

- <u>Study director qualifications</u>. The study director must have the appropriate education, training, and experience to carry out her/his responsibilities. The inspector's review of the facility's personnel records should provide an indication of whether the study director's qualifications are adequate.

- <u>Study director assurances</u>. The inspector should determine whether the study director has assured the following:

- The study protocol, including any changes, is properly signed and approved (see 40 CFR §160.120) and is followed.

- All experimental data, including observations of unexpected test system responses, are recorded and verified.

- Unforeseen events that may affect the quality and integrity of the study are noted and corrective actions taken and documented.

- Test systems are as specified in the protocol.

- All applicable GLP regulations are followed.

- All raw data, documentation, protocols, specimens, and final reports are archived at or before completion of the study.

Problems in these areas indicate that the study director has not performed as required by the GLP regulations.

To determine whether the study director is properly discharging this responsibility, the inspector should obtain some basic information about how the study director operates. In particular, the inspector should determine the level of contact the study director has with the conduct of the study through such information as:

- The relationship of the study director to the study being conducted (i.e., level of involvement)

- Whether the studies are conducted onsite or at another facility

- The level of contact the study director has with any work done at other facilities involved in the study

- The frequency at which the study director reviews the conduct of the study or the collection of data

- The availability of the study director when changes in the conduct of the study are required

- The process by which changes in procedures are authorized.

Also, the inspector should determine if the study director assures that data are archived in a timely manner.

4.3.4 Quality Assurance Unit (QAU) [40 CFR §160.35]

The facility must have a QAU independent from the personnel actually directing and performing the study. The QAU is responsible for monitoring each study to assure management that the facilities, equipment, personnel, methods, practices, records, and controls are in conformance with the GLP regulations. The organizational chart should show that the QAU is independent from the conduct of the study.

The inspector should verify that the QAU is fulfilling its responsibilities with respect to the conduct of regulatory studies, but is not permitted to examine reports of QAU inspection findings and problems, or actions recommended and taken.

The QAU is responsible for the following items:

- Master schedule. The inspector should obtain a copy of the master schedule and check it for conformance with the GLP regulations. The QAU must maintain a copy of the master schedule of all studies conducted at the testing facility, indexed by test substance, and containing information on the test system, nature of the study, the date the study was initiated, the current status of each study, the identity of the sponsor, and the name of the study director.

- Maintain protocols. The QAU must maintain copies of all protocols for which the unit is responsible. The inspector should determine whether all such protocols are readily available, and whether they are up to date.

- Study inspections. The QAU must inspect the study at intervals adequate to assure study integrity and maintain written records of these inspections. The inspector should examine these inspection records to determine whether they are complete, properly signed, and show the following:

    - Date of inspection
    - The study inspected
    - The phase or segment of the study inspected
    - The inspector's name.

    The inspector should determine if the records show adequate phase checks on ongoing studies. The inspector should also determine whether any problems likely to affect study integrity were brought to the attention of the study director and facility management immediately following inspection.

- Reports to management and the study director. The QAU must periodically submit to management and the study director written reports on the status of each study, noting problems and any corrective actions taken. The inspector should determine that notification of QAU findings were provided.

- <u>Deviations from protocols and SOPs</u>. The QAU must assure that no deviations from approved protocols or SOPs have occurred without proper authorization and documentation. The inspector should interview QAU personnel and review QAU reports to determine whether the QAU has detected unauthorized deviations. If the inspector finds any deviations during the inspection, s/he should check to see if they were authorized.

- <u>Review final study report</u>. The QAU must review the final study report to assure that this report accurately describes the methods and SOPs actually used and that the results in the report accurately reflect the raw data collected during the study. While an ongoing study will probably not have reached the stage where the final study report has been written, the inspector may check whether the QAU reviewed the report for another recently completed study.

- <u>Statement with final study report</u>. The inspector should examine a final study report, if one is available, to determine whether the QAU signed a statement in that report specifying the dates inspections were made and findings reported to management and the study director.

- <u>Required records</u>. The QAU must maintain records of the responsibilities and procedures, and the indexing for such records. The inspector should examine such records maintained by the QAU (some of which were described previously in this section) to assure that they are complete, detailed, and accurate. The inspector may also request that facility management certify that inspections are being performed as required in 40 CFR §160.35. (Determine this by checking whether the records indicate periodic inspection showing the date of the inspection, the study inspected, the phase or segment of the study inspected, and the name of the person performing the inspection. Also determine if one or more inspection of a phase or segment of the study was conducted while the study was ongoing.) The quality of the data being recorded should also show compliance with GLP regulations; where this is not the case, the QAU inspection procedures have failed. The inspector may also note whether the QAU appears to have the support of test facility management.

4.4 FACILITIES

The inspector should take particular note of the nature of the facilities provided for the studies during the walk-through, as this provides the best opportunity for assessing their adequacy.

4.4.1 Test Facility, General [40 CFR §160.41]

The inspector should note whether the facility is of suitable size and construction for the proper conduct of the studies. Functions that have the potential to adversely affect the study must be conducted in an area sufficiently separated from other study areas to preclude these effects. The space, bench area, storage, environmental chambers, caging, water baths, water quality and equipment, and other facilities must be appropriate for the type of testing being conducted. The inspector should describe the size and construction of the facility and take photographs, when appropriate.

4.4.2 Test System Care Facilities [40 CFR §160.43]

- <u>Test system separation</u>. The facility must have a sufficient number of rooms or other areas for proper separation of species or test systems, isolation of individual projects, quarantine or isolation of animals or other test systems, and routine or specialized housing of test systems. Areas for plants or aquatic projects must comply with the requirements found at 40 CFR §160.43(a)(1) and (2) as appropriate.

- <u>Biohazardous areas</u>. Separate areas must be provided for tests using biohazardous substances or test systems.

- <u>Areas for diseased test systems</u>. Separate areas must be provided as appropriate for the diagnosis, treatment, and control of test system diseases. The areas must provide effective isolation of test systems known or suspected of being diseased or of being carriers of disease.

- <u>Disposal of materials</u>. The facility must have proper facilities for collection and disposal of contaminated water, soil, or spent materials. Waste from housed animals must be collected and disposed of so as to minimize vermin infestation, odors, disease hazards, and environmental contamination.

- <u>Environmental conditions</u>. The facility must have provisions to regulate environmental conditions (e.g., temperature, photoperiod) as specified in the protocol.

- <u>Test system media</u>. The facility must have an adequate supply of water (for aquatic organisms) or soil of the appropriate composition (for plants) as specified in the protocol.

4.4.3 Test System Supply Facilities [40 CFR §160.45]

The facility must have appropriate storage areas for feed, soil, nutrients, bedding, supplies, and equipment. Storage areas for feed, nutrients, soils, and bedding must be separate from areas where test systems are located, and must be protected against infestation or contamination. Perishable supplies must be appropriately preserved. Test system holding and culturing areas must also be provided (i.e., ponds, culture areas, greenhouse, holding tanks, or fields for aquatic animals or plants, as appropriate).

4.4.4 Facilities for Handling Test, Control, and Reference Substances [40 CFR §160.47]

The facility must have separate areas, as necessary, for the following:

- Receipt and storage of test, control, and reference substances
- Mixing of the test, control, and reference substances with a carrier
- Storage of such mixtures.

Storage areas for test, control, and reference substances and mixtures are to be separate from areas housing the test system. These storage areas (freezers and refrigerators in the pesticide testing facility) must also be adequate to preserve the identity, strength, purity, and stability of the substances and mixtures.

4.4.5 Testing Facility Operations Areas [40 CFR §160.49]

The facility must have adequate separation of testing facility space and other space (e.g., glassware wash area) for the routine and specialized activities of studies.

4.4.6 Specimen and Data Storage Facilities [40 CFR §160.51]

The facility must provide archives for storage and retrieval of raw data and specimens from completed studies and for limiting access to the archives.

4.5 EQUIPMENT

4.5.1 Equipment Design [40 CFR §160.61]

Equipment for use in data operations and for environmental control must be of appropriate design and adequate capacity to function according to the protocol or SOPs. The location of equipment must allow proper operation, inspection, cleaning, and maintenance.

4.5.2 Maintenance and Calibration [40 CFR §160.63]

Equipment used in any study must be adequately inspected, maintained, and calibrated. The inspector should confirm that SOPs adequately describe the use and maintenance of the equipment and the persons responsible for such operations. Written records must be available to document the inspection, maintenance, and calibration and/or standardization of the equipment, as well as any repairs resulting from failure or malfunction. In addition, the facility's SOPs must contain detailed information on the methods, materials, schedules, and the person(s) responsible for the performance of these activities.

The inspector should review individual records to determine compliance with the equipment requirements. The records must provide complete and detailed documentation of inspection, maintenance, cleaning, calibration, and repair.

4.6 TESTING FACILITIES OPERATION

The inspector should review the facility's SOPs as appropriate to assure they are complete, detailed, and meet the requirements of the regulations.

4.6.1 Standard Operating Procedures [40 CFR §160.81]

- <u>SOPs</u>. The facility must have adequate, written SOPs for at least those activities listed in 40 CFR §160.81(b), as appropriate. The inspector should review the testing facility's list of SOPs, if a list exists, or the set of SOPs themselves to determine whether all necessary SOPs are available and have the appropriate signatures. Each testing facility area must have available the SOPs applicable to the procedures performed in the area.

The inspector should review selected SOPs, in detail, to determine if they are adequate. The SOPs selected should be relevant to the study being used as an aid in the compliance review, and the review should include an assessment of whether the SOPs were adhered to or, if not, whether any deviations were properly documented.

- Deviations from SOPs. Deviations from SOPs must be authorized and documented. Significant changes to the SOPs must be authorized in writing by management. The inspector should check for documented deviations and changes and for appropriate signatures on the SOPs. The inspector should also note during the walk-through any instances of the staff not following accepted procedures, including areas such as recording of data; use of safety equipment; calibration, use, or maintenance of equipment; handling of samples; or cleaning of animal quarters. The inspector should check these deviations to see if they have been authorized.

- Historical SOP file. The inspector should check whether the facility maintains a historical file of all SOPs and their revisions, including dates of revisions.

4.6.2 Reagents and Solutions [40 CFR §160.83]

The containers of reagents, solvents, and solutions in testing facility areas must be adequately and appropriately labeled to indicate identity, titer or concentration, storage requirements, and expiration date. Deteriorated or outdated reagents and solutions must not be used. The inspector should note any inconsistencies in labeling observed during the walk-through.

4.6.3 Animal and Other Test System Care [40 CFR §160.90]

The inspector should determine whether SOPs discussing the housing, feeding, handling, and care of animals and other test systems exist and whether they are followed. The SOPs for test system care, as well as actual care practices at the facility, must meet the following requirements:

- Newly received test systems must be isolated until their health status or appropriateness can be determined.

- Test systems must not begin a study while diseased or in any other condition that might interfere with the study. Test systems that contract such a disease or condition during the study must be isolated. The facility must keep documents on diagnosis and treatment of diseased test systems.

- Test systems must be appropriately identified, as required by 40 CFR §160.90(d).

- Test systems of different species, or of the same species but different studies, must be housed separately, unless integrated housing is called for by the protocol.

- Appropriate schedules for cleaning and sanitizing all test system holding areas must be arranged.

- Feed, soil, and water used for the test systems must be analyzed to assure that potential contaminants are not above levels set in the protocol, documentation of such analyses must be maintained.

- Animal cage or pen bedding must be maintained and replaced as necessary to assure animals stay clean and dry.

- Use of pest control materials must be documented; such materials must not interfere with the study.

- Plant and animal test systems must be acclimated to test environmental conditions prior to the study.

The inspector should determine compliance by interviewing study personnel and visiting facility areas where the test system is housed. In addition, the inspector should review SOPs for the housing, feeding, handling, and care of the test system in detail to determine if they are adequate. The SOPs selected should be relevant to the study being used as an aid in the compliance review; the review should include an assessment of whether the SOPs were adhered to or, if not, whether any deviations were properly identified.

4.7 TEST, CONTROL, AND REFERENCE SUBSTANCES

4.7.1 Test, Control, and Reference Substance Characterization [40 CFR §160.105]

The inspector should examine facility records to determine whether the appropriate test, control, and reference substance characterizations have been conducted. The identity, strength, purity, and composition or other characteristics that will appropriately define the test, control, or reference substance must be appropriately documented before the use of the substance in a study. The method of synthesis, fabrication, or derivation of the test substance must be documented and the location of the documentation specified. The solubility of the test substance, if relevant, must be determined before the start of the study. The stability of the test substance must also be periodically determined. The test container(s) must be adequately labeled with the name of the substance, the Chemical Abstracts Service (CAS) number, or code number and batch number, expiration date, and storage conditions. Test substance containers must be retained. For studies lasting more than 4 weeks, a reserve sample of the test substance must also be retained. Finally, the stability of the test substance under the existing storage conditions must be known.

4.7.2 Test, Control, and Reference Substance Handling [40 CFR §160.107]

The inspector, based on observations made during the inspection, should determine whether adequate procedures have been established for handling the test, control, and reference substances used during the conduct of the study. Storage conditions must be adequate; distribution practices must preclude contamination, deterioration, or damage; proper identification of the test substance must be maintained throughout its handling; and a use or distribution log for the test substances must be maintained.

4.7.3 Mixture of Substances with Carriers [40 CFR §160.113]

Appropriate analytical methods must be used to test each test, control, or reference substance mixed with a carrier to determine the following: the uniformity and concentration of the mixture; the solubility of each substance in the mixture prior to the experimental start date, if relevant; and the stability of the mixture before the experimental start date or with each new batch. The expiration dates of any of the ingredients in the mixture must not be exceeded. If a vehicle is used for mixing the test substance with a carrier, evidence must be provided that it did not interfere with the integrity of the test.

4.8 PROTOCOL FOR AND CONDUCT OF A STUDY

4.8.1 Protocol [40 CFR §160.120]

Each study must have an approved written protocol that clearly indicates the objectives of the study and all methods to be used. The inspector should examine the protocol for the ongoing study and look for information on the following areas:

- Description of the objectives and the methods to be used
- Identification of the test, control, and reference substances
- Name and address of the sponsor and test facility
- Proposed experimental start and termination date
- Justification for the selection of the test systems
- Description of the test systems
- Procedure for identification of the test system
- Description of the test design, including bias control methods
- Description of the diet for the test systems
- Route of exposure
- Each dosage level of test, control, or reference substances to be administered
- Type and frequency of tests, analyses, and measurements
- Records to be maintained
- Date of approval of protocol by study director and sponsor
- Dated signature of study director
- Statement of proposed statistical methods to be used
- All changes in or revisions to the protocol, including signatures, reasons, and dates.

4.8.2 Conduct of a Study [40 CFR §160.130]

The study and monitoring of test systems must be conducted in accordance with the protocol. The inspector should be familiar with the content of the protocol for the study. During the inspection, s/he should attempt to document the occurrence of events reported to be already accomplished. The inspector should ask to see either the samples taken or evidence that the samples, observations, or measurements have been taken, handled, analyzed, or processed on schedule as described in the protocol or the appropriate SOP. All samples must be labeled with appropriate information and handled according to the prescribed procedures. The inspector should note anything that was atypical for the type of study being investigated, such as:

- Test vessel size
- Amount of test substance used
- Size, number, or handling of samples used or taken
- Age of test systems
- Unusual timing of observations
- Unexpected trends in biological, chemical, or environmental data collected to date; or distribution or handling of test systems or test substances.

Any inconsistencies in the recorded data should be noted, such as normally sequential events not being recorded in a logical sequence, along with any failure to account for all test systems and test substances obtained for use in the study.

The inspector should determine whether the following requirements are also met:

- <u>Specimens</u>. Specimens must be properly identified by test system, study, nature, and date of collection. This information must be located on the specimen container or accompany the container in a manner that precludes errors in data recording or storage.

- <u>Histopathology</u>. In studies involving histopathology, information from postmortem observations must be available to a pathologist at the time of histopathological review.

- <u>Data recording</u>. Data (unless recorded in automated data systems) must be recorded in ink, dated on the day of entry, and signed or initialed by the recorder. Any change must not obscure the original entry, must indicate the reason for the change, and the person responsible for the change must be identified.

4.9 RECORDS AND REPORTS

4.9.1 Reporting of Study Results [40 CFR §160.185]

In a compliance review, this part of the GLP standards will generally not be involved. The inspector can determine the procedures that are in place to assure the study director either obtains copies of the raw data for processing or a signed report from the staff. There should be data available to support all aspects of the study.

By examining a recent, complete report, the inspector can determine that all pertinent required elements are included in the report, and especially that the report includes a quality assurance statement and statement of compliance or noncompliance for the parts conducted at the site. The inspector should check the completed report for the date of the study director's signature and the date of the signature on the compliance statement.

4.9.2 Storage and Retrieval of Records [40 CFR §160.190]

The facility must have storage areas for all the data, documentation, records, protocols, specimens, and final reports relating to the studies. Correspondence and other documents relating to interpretation and evaluation of data must also be kept. Temporary archives must be available for data being generated by the facility during the ongoing study. A person must be designated as responsible for the archives, and entry to the archives must be limited to authorized personnel. Material retained or referred to in the archives must be indexed for expedient retrieval. The inspector should review the procedures the facility follows to archive all the data for the study and the final report, and determine whether these meet the requirements of 40 CFR §160.190, as appropriate.

4.9.3 Retention of Records [40 CFR §160.195]

The inspector should determine whether records, raw data, and specimens have been retained for the period prescribed by the appropriate regulations. For TSCA GLP studies, records must be kept for at least 10 years following the effective date of the applicable final test rule, except in the case of testing under Section 5 of TSCA, for which records must be kept for at least 5 years following the date of submission of the study results to EPA. For FIFRA GLP studies, records must be kept (1) for the period for which a research or marketing permit is held, if the study has been used to support the application for that permit; (2) for a period of at least 5 years following submission to EPA in support of a research or marketing permit; or (3) in any other case, for a period of at least 2 years following termination, completion, or discontinuation of the study.

4.9.4 Special Considerations for Field Sites

Several practical considerations should be addressed by each member of the inspection team before entering the facility and/or during the inspection. The following are some issues that the team members should be aware of when in the field:

- Climatic conditions. In the field environment, weather conditions may vary from day to day during the inspection. Team members should be able to accommodate the field weather conditions within reason. Appropriate clothing, such as boots, long pants, jackets, and rain gear, should be available if the weather changes. Hats, sunglasses, and sunscreen are additional items that should be available in the field.

If the team wants to observe a critical phase of an ongoing study, such as an application of test substance, they should be prepared for the possibility of an early morning (3 a.m. to 7 a.m.) application window. Team members should understand that early morning weather conditions are oftentimes more favorable to the study.

- Field safety. According to safety statistics, the farming environment is one of the most dangerous working environments in the United States. At the field location, team members will most likely be in close proximity to power take-off equipment, hot manifolds, sharp edges, hydraulic machinery, and a multitude of other types of equipment. To avoid being injured, the team members should keep a safe distance from these types of equipment at all times. If, for some reason, inspection of a piece of operating farm equipment is necessary, extreme caution must be exercised, and appropriate safety questions should be addressed to the equipment operator or other appropriate personnel prior to conducting any equipment inspection. Loose clothing of any type should never be worn in the field. Team members need to remain alert at all times, and should always keep long hair, shirt sleeves, and pant legs away from equipment.

- Indigenous animals. As one may expect, the field environment may be infested with bugs, flies, ants, and other pests. Team members may wish to take insect repellant along on the inspection. (Extreme caution should be exercised in the use of insect repellant [a pesticide] in the vicinity of a pesticide study.) If a team member has a known allergic reaction to an insect bite or sting, the team leader should be informed so that appropriate precautions can be taken. Many times skunks, snakes, rodents, raccoons, and other wildlife may be in the area. Team members should be cautious around these animals and remain at a safe distance from them.

- Field data. Field data may be difficult to obtain at times at remote facilities. These data are often forwarded to the sponsor at the completion of a study. The inspector should contact facility personnel well in advance of the inspection and coordinate with them such details as having data available at the site for auditing. This will help to assure that all documents and inspection issues are covered before the team arrives at the field site.

- Test substance storage. Adequate storage conditions for test substances is commonly a problem at many field locations. The team member who inspects this issue should know the storage stability limits for the test substance (this is often given on the label for registered pesticides, but otherwise should be provided by the study sponsor) and should verify that the limits have not been exceeded. If the stability requirements, such as temperature, have been exceeded, the inspector should be notified and appropriate documentation should be obtained.

- Weather data. The study protocol and/or SOPs usually specify that weather data be collected periodically, especially during the test substance application phase of the study. The auditor should verify that this has been done and should compare site records with the protocol and study report. Any discrepancies should be noted and documented, and the inspector should be advised.

CHAPTER FOUR

- Sampling, sample storage and transfer. These areas are of critical importance to the conduct of most studies. Facility SOPs and the protocol should be carefully reviewed and compared to sampling data to determine if the functions were in fact carried out by facility personnel accordingly. Typically, samples have not been monitored according to protocol or facility SOP requirements, which may then lead to distorted results by the analytical laboratory.

- Test and control plot histories. Most field cooperators have very extensive plot histories (previous crops, cultural practices, and pesticide usage) for their own property. When a contractor/cooperator leases property from a local farmer, these histories may not exist or the cooperator may not have asked for them. The team member should verify the plot histories of all study plots, including the control plot(s), and note any absent or sketchy plot histories.

5.0 Audit Procedures

5.1 Introduction

The purpose for conducting audits of studies submitted to EPA under FIFRA or TSCA is to assure the integrity and reliability of the study data. In evaluating the quality of the study data, the auditor must verify that, where applicable, the data were generated in compliance with GLP regulations (40 CFR Parts 160 and 792, respectively). In addition, the auditor must determine if (1) all of the data have been retained, (2) the study can be reconstructed from the data, and (3) the study findings and conclusions are supported by the raw data, including any data that were not considered while drawing conclusions. The study audit is normally accomplished through the completion of the following activities:

- Review of available raw data, records, and reports
- Interviews with study personnel
- Review of testing facility operations and practices.

5.2 GLP Compliance

In addition to the GLP compliance review (see Chapter 4), the auditor and/or inspector must determine that all studies being audited were conducted in such a manner as to comply with GLP regulations and, if not, document such deficiencies:

- FIFRA. Effective October 16, 1989, 40 CFR Part 160 is applicable to "studies which support or are intended to support applications for research or marketing permits for pesticide products regulated by the EPA. This part is intended to assure the quality and integrity of data submitted pursuant to sections 3, 4, 5, 8, 18, and 24(c) of FIFRA..." A study is defined in 40 CFR §160.3 as "any experiment ... in which a test substance is studied ... to determine or help predict its effects, metabolism, product performance (efficacy studies only as required by 40 CFR §158.640), environmental and chemical fate, persistence and residue, or other characteristics in humans, other living organisms, or media. The term 'study' does not include basic exploratory studies carried out to determine whether a test substance or test method has any potential utility."

  Prior to October 16, 1989, but on or after December 29, 1983, 40 CFR Part 160 was applicable only to "any *in vivo* or *in vitro* experiment in which a test substance is studied ... to determine or help predict its toxicity, metabolism, or other characteristics in humans and domestic animals. The term does not include studies utilizing human subjects or clinical studies or field trials in animals. The term does not include basic exploratory studies carried out to determine whether a test substance has any potential utility or to determine physical or chemical characteristics of a test substance."

- TSCA. Effective September 18, 1989, 40 CFR Part 792 is applicable to "studies relating to health effects, environmental effects, and chemical fate testing. This part is intended to assure the quality and integrity of data submitted pursuant to testing consent agreements and test rules issued under section 4 of TSCA..." A study is defined (§792.3) as "any experiment ... in which a test substance is studied ... to determine or help predict its effects, metabolism, environmental and chemical fate, persistence, or other characteristics in humans, other living organisms, or media. The term 'study' does not include basic exploratory studies carried out to determine whether a test substance or a test method has any potential utility."

Prior to September 18, 1989, but on or after December 29, 1983, 40 CFR Part 792 was applicable only to "studies relating to health effects, environmental effects, and chemical fate testing. This part is intended to assure the quality and integrity of data submitted pursuant to section 4(a) of TSCA." A study is defined (§792.3) as "any *in vivo* or *in vitro* experiment in which a test substance is studied ... to determine or help predict its fate, toxicity, metabolism, or other characteristics in humans, other animals and plants. The term does not include studies utilizing human subjects or clinical studies. The term does not include basic exploratory studies carried out to determine whether a test substance has any potential utility."

Therefore, not all studies that may be audited by EPA were required to be conducted in compliance with GLP standards. Thus, the auditor must first determine which, if any, GLP regulations apply to the study(ies) that s/he will be auditing. This should be done as early in the inspection process as possible. If there is any uncertainty about the applicability of the regulations to a study, the auditor should contact the inspector or the Chief, SSB, LDIAD, for assistance in making the determination.

If it is determined that a study was not required to be performed in accordance with GLP standards, then this portion of the study audit need not be conducted. The FIFRA study audit will then consist primarily of a data reliability review and the determination that all original raw data have been retained for studies involving registered pesticides regulated by 40 CFR §169.2(k), Books and Records.

LDIAD SOP GLP-C-02 describes in detail standard procedures to be used for determining the compliance of studies with respect to the GLP regulations. This SOP was written for nonhealth effects studies, but the basic principles are applicable to all studies. The auditor should be familiar with this SOP before attempting to audit the GLP compliance of a study. Additional guidance and assistance should be available from the inspector.

The following GLP-related topics may be evaluated for each study:

- Compliance statement
- Study director
- Quality assurance role
- Master schedule
- Facility environmental information
- Test system information
- Test chemical information
- Equipment logs

- SOPs
- Personnel qualifications
- Protocols and approved changes or revisions
- Receipt information
- Archives.

If a GLP review is to be conducted as part of the study audit, the auditor should review the study report before entering the facility to ascertain that it contains all applicable elements required by the GLP regulations. Additionally, the auditor should also review the study protocol and deviations for each study during the audit to determine adequacy of, and compliance with, the protocol.

During the review of the data, the auditor should determine if the study was conducted as described in the protocol and its approved changes or revisions, and if documentation is available to demonstrate that the study director was notified of, and approved, any deviations from the signed protocol.

The auditor should also determine if the personnel involved in the study were qualified and adequately trained, and that the facilities and equipment were of appropriate size, design, and capacity to function according to the protocol. To accomplish this, personnel training records, facility floor plans, and equipment maintenance and calibration records for the time that the study was conducted should be reviewed as part of the audit. For detailed information on review of these elements, the auditor should refer to Chapter 4.

In conducting a study audit, the auditor will be reviewing and verifying all the same elements as in the compliance review. However, the SOPs reviewed will be the ones in place at the time of the study, rather than the SOPs currently in place (as in the case of a compliance review).

Study auditors should review the required report elements, described in SOP GLP-C-02 to familiarize his/herself with all the information necessary to complete the audit. This information should be recorded in the auditor's field notebook for reference when writing the report.

5.3 Data Review

A number of different techniques may be used to determine the integrity and reliability of study data and study reconstructability. During the review of the study report and before entering the facility, the auditor should develop the approach that s/he will use.

A useful approach is to follow different aspects of the study (test substance application, specimen analysis) chronologically. For example, in a typical study the auditor can:

CHAPTER FIVE

- Trace the test substance from receipt through characterization (if done by the facility), storage, distribution, and application to the test system, checking all data and records in a systematic fashion.

- Check any data and records of analyses to determine the stability and uniformity of mixtures of test substance with a carrier, such as feed mixtures used in toxicology or metabolism studies.

- Review procedures for sampling and shipment of the specimens to a testing facility for analysis.

- Review records of storage and transfer of the specimens and of receipt, storage, and distribution by the analytical testing facility.

- Review any method development or method validation and audit data for the actual specimen analyses.

During an audit, the auditor needs to determine the percentage and number of data points that should be audited. Those will depend largely on the amount of data and any problems subsequently encountered. For a small study, it is usually desirable to audit most or all of the data points; for a large study, the auditor may be able to assess data quality and integrity after auditing as little as 10 to 15 percent of the data. However, if data quality problems are encountered, the auditor should increase the number and percentage of data points audited to determine how widespread any data quality problems are. In some cases, it may be necessary to audit all data, or all data of a single type, even for very large studies.

The auditor should assure that raw data were properly and promptly recorded, and that study data are accurately reflected in the inspection report. The auditor should be especially careful to review any data generated but, for one reason or another, not incorporated into the final report. The facility should have a valid and defensible reason for rejecting or not including data, especially when the results do not support the study conclusions. A description of all circumstances that may have affected the quality or integrity of the data shall be included in the final report.

Study reconstructability means that sufficient records and raw data are present to allow the auditor to fully trace the procedures, operations, specimen analyses, calculations, and data analysis and evaluation used to produce the final study report. The data and records that are reviewed as part of a study audit should be complete enough to permit this reconstruction. If the study cannot be reconstructed, the auditor should determine whether the data were never recorded or whether data were simply not retained. The inspector should then be advised so that the proper documentation can be obtained.

When deficiencies involving data quality, integrity, or retention are encountered, it is essential that they be adequately documented and that proper evidence be collected to support any potential enforcement action. The auditor should refer to Chapter 6 for detailed procedures for collecting documentary samples.

If there is any doubt regarding documentation of potential GLP violations or other data problems, the inspector should be consulted.

5.4 COMMON DEFICIENCIES

Common deficiencies in the study report include, but are not limited to, the following:

- Lack of study director's signature

- Failure to provide the name and address of each unit (e.g., all field sites, processing sites and testing facility sites) of the facility performing the study

- Inadequate or missing quality assurance statement

- Inadequate or missing compliance statement

- Failure to provide the name of the study director and/or names of other scientists, professionals, and supervisory personnel involved in the study

- Lack of the signed and dated reports of individual scientists or other professionals involved in the study, including those who conducted an analysis or evaluation of data or specimens from the study after data generation was completed.

Common deficiencies in the study protocol include, but are not limited to, the following:

- Incomplete description of experimental design (e.g., failure to disclose or reference all methodologies)

- Failure to adequately execute changes or revisions to, or deviations from, the signed protocol

- Failure to identify the study director or lack of the study director's signature.

5.5 GUIDELINES FOR AUDITING STUDIES

Auditors conducting data audits should have and be familiar with all appropriate GLP program SOPs. These SOPs are intended to provide specific guidance. A list of SOPs is give on page 5-6.

5.6 SOP REFERENCE LIST

The following is a list of the SOPs that EPA has developed to provide guidance in conducting compliance reviews and study audits. All study auditors are encouraged to familiarize themselves with the appropriate SOPs before conducting an audit. This is the list of SOPs as of the effective date of this Manual. Inspectors and auditors have a responsibility to keep an updated, current list of relevant SOPs.

CHAPTER FIVE

GLP-C-xx

GLP-C-01   Conducting a Field Site Compliance Inspection. 10/01/90

GLP-C-02   Determining Compliance of Audited Studies with GLP Requirements. 10/01/90

GLP-D-xx

GLP-DA-01   Auditing Field Studies (Analytical Chemistry). 02/01/91

GLP-DA-04   Auditing Residue and Environmental Fate Studies (Field Portions). 01/01/91

GLP-DA-06   Auditing Efficacy Studies. 06/15/91

GLP-DA-07   Auditing Nature and Magnitude of the Residue in Livestock Studies (Biology Portions). 06/15/91

GLP-S-xx

GLP-S-01   Preparation of Standard Operating Procedures. 10/01/90

GLP-S-02   Evidence Requirements for Documenting GLP Standards and Study Audit Deficiencies. 01/15/91

GLP-S-03   Format of GLP Inspection/Data Audit Summary Report. 06/16/91 (To be replaced by GLP-S-07)

GLP-S-04   Format of GLP Inspection Comprehensive Report. 04/01/91

GLP-S-05   Glossary of GLP Terms. 01/01/91

GLP-S-06   GLP Inspection Review Committee (GRC) Procedures. 12/01/90

GLP-S-07   Format of A Brief GLP Inspection Report. **(Pending)**

6.0 Post-Inspection Activities

6.1 Introduction

EPA inspectors conduct compliance inspections to verify that the regulated community is complying with the FIFRA and TSCA GLP regulations. In addition, regulatory studies that have been submitted (or that are intended for submission) to EPA are audited to verify that reported findings and conclusions are consistent with the raw data and other supporting records, reports, and correspondence for the studies.

The effectiveness of any GLP compliance inspection depends on many factors, including the thoroughness of the inspection, the evidence collected by the inspector, and the cooperation of the facility being inspected. Also critical to the success of the inspection are three steps that follow the inspection:

- Conducting necessary followup activities (Section 6.2) at the conclusion of the inspection
- Ensuring the proper collection and accountability of all evidentiary material (Section 6.3)
- Preparing the inspection report (Section 6.4).

The sections that follow include a discussion of followup activities, evidence collection, and report preparation. Followup activities assure that any outstanding documents pertaining to the facility and the inspection are obtained as soon as possible after the inspection for evaluation and inclusion in the inspection report. All GLP deviations and other potential violations must be properly documented so that enforcement and regulatory actions can be successfully initiated and prosecuted. The primary function of the inspection report is to record the events and observations of the inspector, serving as a basis for any enforcement or regulatory decisions will be based.

6.2 Followup

At the conclusion of the onsite phase of a GLP compliance inspection at a test site, additional activities will be required, these include complying with the instructions to the inspector that appear on the Investigation Request, acquiring additional information and preparation of the inspection report. Preparation of the inspection report is most important since this will serve as the basis for any enforcement action. However, there may also be a need to acquire additional information such as copies of documents and/or written statements either from the inspected test site or another facility having information or records pertaining to studies audited or reviewed at the site of the inspection.

The purpose of this discussion is to provide guidance to the inspector on the acquisition of additional evidentiary materials to:

- Adequately support a potential enforcement action
- Provide justification for additional inspections at other facilities
- Satisfactorily resolve issues left unanswered during a test site inspection.

CHAPTER SIX

Study auditors are obligated to become familiar with the procedures described herein to assure consistency throughout the GLP inspection program. The basic procedures described are the same for FIFRA and TSCA, except that each statute's particular CBI procedures must be followed, as appropriate.

6.2.1 Followup Information from an Inspected Facility

At the conclusion of the onsite phase of a GLP inspection, an inspected facility may need to provide additional documents, information, or other evidentiary materials to allow an inspector to complete the inspection report. Examples of such documents may include, but are not limited to:

- Signed statements from facility officials regarding GLP or data issues or circumstances related to inspection findings

- Copies of additional pertinent records, data, correspondence, or other documents that are located by the facility after the inspection team departs

- Additional document copies or information requested by the inspector after the inspection when further review of collected information or records by a member of the inspection team indicates such information is needed

- Copies of data, records, or other documents routinely available from another facility of the same company (e.g., test, control, or reference substance characterization data; stability information; test system source records)

- Voluntary written statements by staff of the inspected facility in response to findings and recommendations presented by the inspection team during the closing conference.

The need for these additional documents may arise (1) from discussions or agreements during the closing conference; (2) as a result of a request from the inspector after return to his/her office; or (3) from the test facility locating additional materials on its own initiative. In the latter two cases, submission of additional materials will usually be preceded by a telephone call between the inspector and the facility regarding the nature of the materials and the most appropriate means of identification and transmittal.

Regardless of the situation, the following procedures must be considered for any followup information or document copies submitted or requested:

- A reasonable deadline should be agreed upon for any additional information, written statements, or document copies that the inspector requests from the test facility at the closing conference or in a request letter.

- Subsequent to the onsite phase of the inspection, any request by the inspector for additional information, document copies, or written statements may be made verbally or in writing, depending on the significance or complexity of the material requested. The written request may be made by the inspector, a staff attorney, or senior program official as local procedural protocol dictates. The auditor, however, should never contact an inspected facility directly to acquire additional information or documents.

- All requested documents, information, or written statements are to be transmitted directly to the lead inspector; it will be the inspector's responsibility to assure that the master file of exhibits is maintained and that appropriate photocopies are distributed to the auditors, to LDIAD, or included in the inspection report.

- If photocopies of data (raw or transcribed), correspondence, records, or other followup materials are to be submitted, the facility should certify, either individually on each copy or via signed statement, that such documents are exact or true copies.

- If copies of the followup documents represent raw data as defined by 40 CFR § 160.3 (FIFRA) or 40 CFR §792.3 (TSCA), a responsible official from the test facility should submit a signed statement as to where the original documents were located, particularly if they were required to have been archived and did not appear to be available in the archives when reviewed during the study audit portion of the inspection.

- Receipt of any documents or other materials received from an inspected test site or other facility subsequent to the onsite phase should be acknowledged by the inspector either with an additional FIFRA or TSCA receipt form (see Section 2.8) or by a short letter identifying the materials received.

- In general, the materials submitted by the test facility, whether data or record copies, written statements, or other documents, should be handled according to the evidence requirements given in Section 6.3.

An example of a request letter is given in Appendix C. In addition to providing details regarding the materials or information requested, an ability to claim such information as FIFRA or TSCA CBI should be discussed, as appropriate.

The inspector should include all information gathered after the completion of the inspection in the inspection report if it supports and documents the inspection report's findings. All information gathered during or after the completion of the inspection not included in the inspection report should be included in the appropriate evidentiary file.

6.2.2 Followup Information from a Non-Inspected Facility

Occasionally, the inspector may need to contact a facility owned or operated by another company to gather additional information or documents related to a particular FIFRA or TSCA GLP inspection. It may or may not be a test site. Usually, the second facility will be a test site or office associated with the sponsor or registrant, although it also could be a study management firm, another contract facility or nonprofit research institution. The need for additional information or documents usually relates to test, control, or reference substance characterization or stability. However, issues related to specimen analysis and evaluation may also require additional supporting information or raw data copies, particularly if there are specialty analyses involved or specimen stability is in question.

Although the primary test site (the inspected facility) may volunteer to acquire the needed materials or information for the inspector, it is recommended that the inspector communicate directly with the other facility. After the inspector returns to the office, the second facility should be contacted in writing regarding the need for document copies or information. Depending upon the followup procedures, the letter may be sent by the inspector, a staff attorney, or other senior program official. The person contacted at the second facility should be the study director or management official identified in the study report or as identified by officials at the second facility. The inspector should contact the facility in writing to request the needed documents or information; however, the inspector may wish to telephone facility personnel first to explain the situation and advise them that the official written request is being sent.

If the requested materials are incomplete or have significant GLP or other apparent reliability problems, an official GLP standards inspection may be warranted at the secondary facility to fully document a suspected violation or other problem. In such cases, the inspector's supervisor should be consulted and an inspection arranged through the LDIAD office. The details regarding the scope of the inspection and the specific personnel involved will be contingent on the circumstances involved and the facility's location.

6.3 THE INSPECTION REPORT

The purpose of an inspection report is to present a complete and factual record of the inspection process from opening conference, through the inspection, to closing conference. The report should contain enough information about the facility and the inspection (as well as observations made during the inspection) to enable Case Development Officers (CDOs) to make enforcement decisions pertaining to the inspected facility and to develop a case, as necessary.

The inspector should prepare the inspection report as soon as possible following the inspection. EPA recommends that the report be completed within 60 days of the inspection; however, the actual amount of time will depend on obtaining any additional required information in a timely fashion. This timeframe should allow the inspector sufficient time to conduct necessary follow-up and to append to the report (and mention in the narrative) any data obtained during follow-up.

As the inspector prepares the report, s/he should have the following objectives in mind:

- To include in the report all of its basic elements, ensuring that the report not only contains copies of relevant forms and documents as appendices, but that the narrative component of the report references those forms and documents

- To substantiate with as much evidence as possible each potential violation of FIFRA or TSCA GLP Standards violations, again ensuring that any documents and/or photographs are not only appended to the report, but are referenced in the narrative component of the report. (This is necessary so that CDOs know how the data relates to the inspection.)

- To write the report in clear and concise language

- To present factual and accurate information pertaining to all steps in the inspection process from opening to closing conference and follow-up

- To make only those observations that are based on firsthand knowledge of the facility since enforcement personnel must be able to depend on the accuracy of all information

- To include only information that is relevant to the facility and its compliance with FIFRA or TSCA. (Irrelevant facts can interfere with enforcement decisions.)

The inspection report should not, under any circumstances, include the inspector's conclusions regarding compliance or noncompliance. Conclusions should be contained in a separate memorandum or other format that is clearly separate from the inspection report. The reason for this is that in an enforcement case, the entire inspection report is subject to discovery by the opposing side. If conclusions of law or opinions are in the report, it may weaken the inspector's credibility by suggesting bias. In addition, the inspector may have been wrong about one or more violations and EPA did not pursue them. This would be revealed through discovery and would again weaken the inspector's credibility. A separate memorandum of findings or conclusions will usually be protected from discovery based on attorney-client privilege or another exception rule.

6.4.1 Elements of the Inspection Report

There are certain elements that should be contained in each inspection report to ensure that necessary information is not inadvertently overlooked. The report should always contain enough information so the reader can determine:

- Specific reason for the inspection
- Participants in the inspection
- Compliance with all required notices, receipts, and other legal requirements
- Actions taken (and chronology)
- Statements, records, physical samples, and other evidence obtained
- Observations made

The inspection report should be a concise and chronological account of observations made and activities undertaken during the inspection, from opening conference to closing conference and follow-up. The field logbook and/or an inspection checklist (if used) are useful tools for developing the narrative. These tools can help the inspector recall and include in the narrative important details concerning the inspection. The inspector should also include the reason for the inspection, any relevant historical information, and any knowledge of prior violations obtained during the pre-inspection process.

CHAPTER SIX

Administrative Exhibits

Exhibits to the inspection report should include all evidence, including affidavits, statements, drawings and maps, mechanical recordings, printed matter, and photographs, that supports the observations made during the inspection (and which should be described in the report narrative, as appropriate). The inspector should prepare an index of exhibits listing the name and the location of each exhibit. This index should precede the exhibits and serve as a reference for enforcement personnel.

There are several forms pertaining to the inspection that should be labeled as exhibits and appended to the end of the inspection report. The most important of these are the forms relating to the FIFRA or TSCA inspection. They should be labeled and attached to the report as follows:

- FIFRA or TSCA Notice of Inspection
- Receipt for Samples and Documents

6.4.2 Inspection Checklists

Inspection checklists are considered to be an extension of the field notebook and are designed to collect standard, reviewable information about an inspection. These forms, however, are only one aspect of the full inspection and are not considered sufficient documentation for the inspection by themselves. They should be completed during the course of an inspection and simply function as guides to ensure that all basic data are collected. If individual items on the forms need clarification or elaboration, the inspector should record it in the field logbook.

6.4.3 CBI Considerations

Some or all of the data gathered during the inspection may be confidential business information (CBI), if properly claimed as such by the facility. Otherwise, the report may be released to the public in response to a Freedom of Information Act (FOIA) Request, unless the report falls under a FOIA exemption. Therefore, if the inspection report contains CBI, those portions of the inspection report must be treated in accordance with FIFRA CBI procedures. However, the inspector may refer to CBI material in general terms (e.g., by a reference number assigned by the inspector) so that the report need not be treated as CBI.

6.4.4 Practical Tips for Report Preparation and Writing

The style of inspection reports should be clear, concise, accurate, factual, fair, complete, and logical. The report must be written to eliminate the possibility of erroneous conclusions, inferences, or interpretations. It will become part of the permanent records for the facility, along with the inspector's field logbook, samples, formal statements, photographs and other pieces of evidence. A well-written report will serve as a summary of these other records.

In general, four rules apply to preparation of good inspection reports:

- Write what the reader needs to read, not what you need to write.

- <u>Write to express, not to impress</u>. Only facts and evidence that are relevant to the compliance situation should be included.

- <u>Keep the report simple</u>. Complicated matters should be organized and stated in simple, direct terms.

- <u>Keep the reader in mind</u>. Writing, language, and terms used should be familiar to the reader.

Keeping these three rules in mind, these basic steps should be followed when preparing to write the inspection report:

- <u>Review the information</u>. As the first step, all information gathered during the inspection should be collected and reviewed, including inspection report forms and checklists. The inspector should then review the information for relevance and completeness. If gaps are identified, follow-up telephone calls can be made or, if necessary, a follow-up inspection can be conducted.

- <u>Organize the material</u>. There are several different methods that can be used for organizing the inspection data. Whatever the method, the material should be presented in a logical, comprehensive manner and organized so it can be easily understood.

- <u>Reference accompanying material</u>. All evidence (e.g., copies of records, analytical results, photographs) that accompany the report should be clearly referenced so the reader can locate them easily. All support documents should be checked for clarity prior to writing.

In writing the report, the procedures used in, and the findings resulting from, the evidence-gathering process should be recorded in a factual manner. The report should refer to routine procedures and practices used, and describe in detail the facts relating to potential violations and discrepancies, but not, as emphasized above, suggestions or conclusions that there may be or are potential violations. The inspector should use the field logbook as an aid to writing the report.

A well-written, effective inspection report has several essential characteristics. By keeping these characteristics in mind when writing, the end result should be a report that provides sufficient data for proper enforcement decisions. The following characteristics, individually, will not ensure a good report, but when addressed together throughout the complete report, will lead to an effective document that can be used as the basis for enforcement actions:

- <u>Fairness</u>. The reports must be entirely objective, unbiased, and unemotional. Distortion, rumors or gossip, or offensive remarks or language should be avoided.

- **Accuracy**. The information should be stated precisely and accurately in plain language. The facts should be presented so clearly that there is no need for conclusions or interpretations.

- **Completeness**. All information that is relevant should be included. Completeness implies that all the known facts and details have been reported, either in the text or in an exhibit. The report should be tested to ensure that it answers the questions "who, what, how, when, where, and why" related to the compliance situation:

    - On first mention, all individuals should be called by their first, middle, and last names
    - Clearly indicate what happened or how it happened
    - Identify the location of the occurrence as a definite place
    - State why a situation is particularly significant with respect to violations.

- **Sources**. The sources of evidence should always be provided. The report should be interview-oriented (i.e., report statements made by interviewees).

- **Conciseness**. Elaborate or unessential information should not be included. Sentences, paragraphs, and tables should be as short as possible.

- **Clarity**. The inspector should minimize the possibility of misinterpretations. Thoughts should be arranged logically and convey the desired message.

- **Organization**. The inspector should organize the report with a logical and coherent order in the presentation of facts.

Appendix A

Completed Notification Letter

UNITED STATES ENVIRONMENTAL PROTECTION AGENCY
WASHINGTON, D.C. 20460

OFFICE OF
PREVENTION, PESTICIDES AND
TOXIC SUBSTANCES

<u>FAX AND EXPRESS MAIL</u>
<u>CONFIRMATION OF RECEIPT REQUESTED</u>

Ms. Ann L. Donargo
Sample Test Facility 000123
Main St.
Alexandria, VA 22314

Dear Ann L. Donargo:

This will inform you that the United States Environmental Protection Agency (EPA) will conduct a Good Laboratory Practice (GLP) inspection at your facility under the Federal Insecticide Fungicide and Rodenticide Act (FIFRA).

The inspection will be conducted on February 14-17, 1987. The inspection will be led by Arunas K. Draugelis. The inspection team will review your facility's compliance status with the EPA FIFRA GLP regulations at 40 Code of Federal Regulations (CFR), Part 160 and will audit those aspects of the studies listed in Attachment I that were performed by Sample Test Facility 000123.

In addition, the inspection team will choose one or more completed or ongoing studies from your Master Schedule for audit.

The purpose of the study audits is to validate data in final reports which have been presented to the EPA in support of a registration or marketing petition under FIFRA.

The purpose of the compliance review is to determine that the GLP regulations of FIFRA are being observed in your testing facility's current procedures and practices for pertinent studies being conducted.

Please note that under FIFRA GLP regulations at 40 CFR 160.15(b) EPA will not consider reliable for purposes of supporting a FIFRA application for a research or marketing permit, any data developed by a testing facility that refuses to permit inspection.

To successfully conduct our inspection, we request that the following matters be addressed prior to our arrival at Sample Test Facility 000123

Recycled/Recyclable
Printed with Soy/Canola Ink on paper that
contains at least 50% recycled fiber

Please make available suitable space for the team. Please have available and in good order all original data needed to verify the final report of each study, along with full copies of the protocol (including protocol amendments) and all reports submitted by your facility to the study sponsors. All current personnel who were associated with these studies should be available for discussion with members of the team as necessary. The inspection team will need for review copies of all Standard Operating Procedures (SOP) documents in use at the time of study.

We will require very specific information at your facility regarding the analytical reference standards (reference substance). This includes, but is not necessarily limited to, the source and lot number, analysis for purity and identification record of receipt and storage, test substance inventory logs and custodial procedures for each reference substance. Records and data should also be available to document the synthesis, radiochemical purity and specific activity of any radio-labeled test or reference substance used at your facility for the conduct of the studies being audited.

In addition, where applicable, please have available any data generated at your facility to verify or validate methodology, for quality control, to establish storage stability, or other related and pertinent analysis.

Please obtain from the sponsor a statement indicating the origin of the test substance, namely, if it was sampled from the batch for contemporary commercial use or was synthesized or manufactured for the specific study for which the raw data are being audited. In either case, chemistry data also include all data to prove the identity and purity of the test substance, the identity of any and all impurities detected by the sponsor or manufacturer, and data to prove storage stability of the test substance during the lifetime of the study.

If there are any questions arising from this notice please feel free to call Francisca Liem, Chief, Scientific Support Branch directly. Under ordinary conditions the dates selected for the inspection will not be changed. Ms. Liem may be reached by telephone during regular hours at (703) 308-8333 or by fax at (703) 308-8285.

Sincerely,

David L. Dull, Director
Laboratory Data Integrity
  Assurance Division
Office of Compliance Monitoring

Appendix B

Investigation Request

UNITED STATES ENVIRONMENTAL PROTECTION AGENCY
WASHINGTON, D.C. 20460

OFFICE OF
PREVENTION, PESTICIDES AND
TOXIC SUBSTANCES

MEMORANDUM

SUBJECT: Investigation Request: Sample Test Facility 000123

FROM: David L. Dull, Director
Laboratory Data Integrity Assurance Division

TO: Rick Dreisch, Chief
Central Regional Laboratory
Annapolis, MD
Region III

It is requested that a Good Laboratory Practice (GLP) Inspection be conducted as described below:

Statute: FIFRA

Type: Neutral Scheme

Dates: February 14-17, 1987

Facility: Sample Test Facility 000123
Main Street
Alexandria, VA 22314

Contact: Ann L. Donargo   Phone: (703) 555-1212

Lab/PDMS No.: 000123   Investigation ID: 8720001231

It is confirmed that the inspector assigned will be:

Arunas K. Draugelis

The inspector should contact the other participants listed below to coordinate travel and accommodation arrangements:

Andrea Blaschka, HQ/OPPT
Arnold Cytryn, FDA

<u>GLP compliance review:</u>

An important activity during this visit will be the GLP compliance review, which should focus on ongoing studies.

Recycled/Recyclable
Printed with Soy/Canola Ink on paper that
contains at least 50% recycled fiber

Study audit:

EPA inspectors will conduct study audits. The study reports listed in Attachment I are eligible for audit.

In addition, a completed study will be selected from the master schedule for audit.

Inspection Coordination:

The inspector will deal with all administrative aspects of this inspection, and will prepare the inspection report.

Reporting:

An Inspection Report of this inspection prepared as per SOP No. GLP-S-03 is due at the office of the Chief of Scientific Support Branch (SSB), Laboratory Data Integrity Assurance Division (LDIAD) **within four weeks after the conclusion of the inspection itself.** **The GLP compliance statement, final study report pages where the study dates are indicated and the cover page of the final study report should be submitted as part of the Summary Report.**

Notification of the Testing Facility:

This office will contact the testing facility in advance of the inspection date. The Inspector may contact the testing facility as needed after the notification is made.

Please feel free to call Francisca Liem (703-308-8333) if we can be of further assistance.

IMPORTANT NOTE TO INSPECTOR

At the conclusion of the inspection the inspector must complete Attachment I to indicate any change in facility, address, contact person, dates, participants, studies audited, ongoing and/or completed studies selected at the testing facility. This form is critical to LDIAD's ability to maintain complete records and generate complete reports of program activities. Please sign and mail to Chief, PSCRB at LDIAD as soon as you return to your home base.

cc: Francisca Liem
    Robert Zisa
    Arunas K. Draugelis
    Andrea Blaschka
    Arnda Cytryn

Attachment I

Lab.   Inspection dates

Address

Telephone No.

Contact person

Type of Inspection

Test substance	Study	Lab. Project No.	OECD Code	Audit Completed Y/N

Ongoing study/studies selected: yes / no

Additional completed study/studies selected: yes / no

Comments:

Inspector:_____

Print name _____

Date:_____

Appendix C

Example Of Request For Further Information Letter

UNITED STATES ENVIRONMENTAL PROTECTION AGENCY
WASHINGTON, D.C. 20460

OFFICE OF
PREVENTION, PESTICIDES AND
TOXIC SUBSTANCES

<u>CERTIFIED MAIL</u>
<u>RETURN RECEIPT REQUESTED</u>

Dr. James Q. Alchemist
Senior Research Chemist
D & H Chemicals, LTD.
Post Office Box 9999
New York, NY 22222

Dear Dr. Alchemist

On February 14-17, 1987, Arunas K. Draugelis, an inspector of the United States Environmental Protection Agency (EPA) conducted a Federal Insecticide, Fungicide and Rodenticide Act (FIFRA) Good Laboratory Practice Standards (GLPS) inspection at Sample Test Facility 000123, Main Street, Alexandria, VA 22314. As part of this inspection Dr. Draugelis conducted an audit of the following completed studies:

Chronic EFFECT OF AC 243,997 to the Water Flea (Daphina Magno) in a 21-day Flow-through Exposure.

Testing of AC 243,997 through FDA Multi-Residue Protocols A through E

These studies were submitted to EPA by D & H Chemicals, LTD. in partial support of a pesticide registration for AC 243,997 under the Federal Insecticide, Fungicide, and Rodenticide Act (FIFRA).

During Dr. Draugelis review with the Sample Test Facility 000123 staff, some raw data could not be located at the facility. The raw data regarding the analytical identification, characterization and stability of AC 243,997 were missing. Please send certified photocopies of the following information for the test substance to Francisca Liem, Chief of the Scientific Support Branch at the address specified in this letter:

Documentation substantiating the identity, purity and characteristics of AC 243,997. Such documentation may include true copies of lab notebook entries, GC or HPLC chromatogram, IR, NMR, or mass spectra data.

If the data are not in your possession or not known to be extant elsewhere, provide Ms. Liem with a written statement from you or another responsible management official attesting to that fact. According to FIFRA Books and Records regulations, 40 CFR Section 169.2 (k), all underlying raw data for the test reports submitted to the Agency shall be retained by the producer for as long as the registration is valid.

Pursuant to regulations appearing at 40 CFR Part 2, Subpart B, and specifically, Section 2.307, you are entitled to claim any or all of the information provided to EPA as confidential business information. If you do not assert a confidentiality claim, the information may be made available to the public without further notice. Such information can be disclosed by EPA only in accordance with the procedures set forth in the regulations (cited above). Any such claim for confidentiality must conform to the requirements set forth in 40 CFR Section 2.203 (b).

Please provide the information directly to Ms. Liem by December 17, 1987. If the data are not provided by this date, we will consider it missing data according to the FIFRA Books and Records regulations, 40 CFR Section 169.2 (k). If you have specific questions regarding this request Ms. Liem may be reached at:

> Environmental Protection Agency
> Office of Compliance Monitoring (7204W)
> 401 M Street, SW
> Washington, D.C. 20460
> Phone: (703) 308-8333

> Sincerely,

> David L. Dull, Director
> Laboratory Data Integrity
> Assurance Division

cc: Ann Donargo (Sample Test Facility 000123)

Appendix D

GLP FIFRA Compliance Checklist

UNITED STATES ENVIRONMENTAL PROTECTION AGENCY
GLP COMPLIANCE INSPECTION CHECKLIST

PART I — GENERAL

GLP FACILITY INSPECTION

Name: _____   Date: _____

Address: _____   Insp. No. _____

City: _____ State: _____ Zip: _____

Phone No.: _____ Contact Person: _____

FACILITY INFORMATION:

Is this facility:

    a sponsor lab? _____
    a contractor lab? _____
    a management company? _____

What types of studies are conducted here (i.e., toxicology, chemical analysis, field) _____

PREINSPECTION REVIEW: (Obtained from Regional Office Files)

Date(s) of Previous EPA Inspection(s): _____

Previous Findings:
_____
_____
_____

REASON FOR INSPECTION:
The purpose of this inspection is to determine if the facility is in compliance with the requirements of FIFRA, codified in 40 CFR Part 160.

☐ Randomly Selected Neutral Inspection
☐ Selected for Cause
    ☐ Referral from _____
    ☐ Other: (Specify) _____

FIFRA GLP INSPECTION CHECKLIST
General Information
Revised 9/93

Laboratory: _____ Insp. Init.: _____ Date: _____

Comments (Please refer to subpart, section, or page numbers):

_____
_____
_____
_____
_____
_____
_____
_____
_____
_____
_____
_____
_____
_____
_____
_____
_____
_____
_____
_____
_____
_____
_____
_____
_____
_____
_____
_____
_____
_____
_____
_____

FIFRA GLP INSPECTION CHECKLIST
Data Audit Review
Revised 9/93

Laboratory: _____ Insp. Init.: _____ Date: _____

OPENING CONFERENCE

PRELIMINARY INFORMATION

1. Laboratory personnel present and interviewed:

   Name: _____ Title: _____
   Name: _____ Title: _____
   Name: _____ Title: _____
   Name: _____ Title: _____
   Name: _____ Title: _____

2. EPA inspector accompanied:

   Name: _____ Agency: _____
   Name: _____ Agency: _____

3. Credentials presented to: _____

4. "Notice of Inspection" signed by laboratory official and copy provided to official?   ☐ Yes   ☐ No

5. Was a GLP Compliance Review conducted?   ☐ Yes   ☐ No
   If so, complete Form I.

6. Was a data audit (or audits) conducted?   ☐ Yes   ☐ No
   If so, complete Form II for each study audited.
   List of studies audited:
   _____
   _____
   _____

CLOSING CONFERENCE (to be completed at conclusion of the inspection)

A. Date: _____ Time: _____ Where conducted: _____
   Facility Representative(s) Present:
   Name: _____ Title: _____
   Name: _____ Title: _____
   Name: _____ Title: _____

FIFRA GLP INSPECTION CHECKLIST
General Information
Revised 9/93

Laboratory: _____ Insp. Init.: _____ Date: _____

Comments (Please refer to subpart, section, or page numbers):

_____
_____
_____
_____
_____
_____
_____
_____
_____
_____
_____
_____
_____
_____
_____
_____
_____
_____
_____
_____
_____
_____
_____
_____
_____
_____
_____
_____
_____
_____
_____

FIFRA GLP INSPECTION CHECKLIST
Data Audit Review
Revised 9/93

Laboratory: _____ Insp. Init.: _____ Date: _____

CLOSING CONFERENCE (to be completed at conclusion of the inspection)

B. Were facility officials provided copies of:

☐ Receipt for Samples and Documents ☐ Inspection Confidentiality Notice
☐ Updated Regulations/Guidances ☐ Declaration of Confidential Business Information

C. Were any documents, records, etc. requested from the facility? ☐ yes ☐ no
(If yes, include the list of information requested, and when it is due to be sent)

D. Does the inspector need to conduct any further follow-up activities? ☐ yes ☐ no
(If yes, please attach an explanation of what must be done, and a projected schedule for the completion of all follow-up activities.)

Inspector's Signature: _____ Date of Signature: _____

NOTES:

FIFRA GLP INSPECTION CHECKLIST
General Information
Revised 9/93

Laboratory: _____   Insp. Init.: _____   Date: _____

Comments (Please refer to subpart, section, or page numbers):

_____
_____
_____
_____
_____
_____
_____
_____
_____
_____
_____
_____
_____
_____
_____
_____
_____
_____
_____
_____
_____
_____
_____
_____
_____
_____
_____
_____
_____
_____
_____
_____

FIFRA GLP INSPECTION CHECKLIST
Data Audit Review
Revised 9/93

Laboratory: _____ Insp. Init.: _____ Date: _____

PART II — GLP COMPLIANCE REVIEW CHECKLIST

FORM I — GLP COMPLIANCE REVIEW

Were any ongoing studies available? Please complete this form for each ongoing study selected.

Study selected for review:

    Test substance: _____

    Study title: _____

    Lab ID No.: _____

    Sponsor (name and address): _____

    Study director: _____

    Study initiation date: _____

    Proposed completion date: _____

GENERAL INSTRUCTIONS/INFORMATION
1. For any "No" answers, provide explanation.
2. Remarks can be continued in the "Comments" section on the back of each page.
3. Place a line through any item missing. For example, "...name/signature..."

SUBPART A — GENERAL PROVISIONS	YES	NO	N/A	REMARKS
§160.10 Applicability to studies performed under grants and contracts				
Has laboratory, contractor, or grantee been informed that their services must be conducted in compliance with 40 CFR Part 160?				

SUBPART B — ORGANIZATION & PERSONNEL	YES	NO	N/A	REMARKS
§160.29 Personnel				
(a) Are training, education, and experience adequate?				
(b) Are training and experience records available?				
(c) Is the number of personnel adequate?				
(d) Are personnel health and sanitation precautions being followed?				
(e) Is appropriate clothing available and worn as needed?				
(f) Are any personnel ill to the extent that they have an adverse effect on the study?				
- If so, are they excluded from direct contact with test systems and substances?				

FIFRA GLP INSPECTION CHECKLIST
Compliance Review
Revised 9/93

Laboratory: _____ Insp. Init.: _____ Date: _____

Comments (Please refer to subpart, section, or page numbers):

_____
_____
_____
_____
_____
_____
_____
_____
_____
_____
_____
_____
_____
_____
_____
_____
_____
_____
_____
_____
_____
_____
_____
_____
_____
_____
_____
_____
_____
_____
_____

FIFRA GLP INSPECTION CHECKLIST
Data Audit Review
Revised 9/93

Laboratory: _____ Insp. Init.: _____ Date: _____

FORM I — GLP COMPLIANCE REVIEW (Continued)

SUBPART B — ORGANIZATION & PERSONNEL	YES	NO	N/A	REMARKS
§160.31 Testing facility management				
(a) Was a study director designated for ongoing study prior to study initiation?				
(b) Has the study director been replaced?				
If so, was this done promptly?				
(c) Is a quality assurance unit in place?				
(d) Are personnel, resources, facilities, equipment, materials, and methodologies available as scheduled?				
(e) Do personnel clearly understand the functions they are to perform?				
(f) Have deviations in the study been communicated to the study director, and have corrective actions been taken and documented?				
§160.33 Study director				
Does the study director have adequate education, training, and experience?				
Is s/he familiar with all aspects of the study?				
Does the study director understand that his/her responsibilities include the following assurances:				
(a) The protocol, including any change, is approved and followed?				
(b) All experimental data are accurately recorded and verified?				
(c) Unforeseen circumstances have been noted and corrective actions taken and documented?				
(d) Test systems are as specified in the protocol?				
(e) All GLPs are followed?				
(f) All data, as required, were transferred to the archives?				
§160.35 Quality Assurance Unit				
(a) Was a separate and independent QAU in place at the time of the study?				
(b) Did the QAU:				
(1) Maintain a complete copy of the master schedule indexed by test substance? (The required elements include the test substance, test system, nature of study, date initiated, current status, identity of sponsor, name of study director.)				
(2) Maintain copies of protocols?				

FIFRA GLP INSPECTION CHECKLIST
Compliance Review
Revised 9/93

Laboratory: _____ Insp. Init.: _____ Date: _____

Comments (Please refer to subpart, section, or page numbers):

_____
_____
_____
_____
_____
_____
_____
_____
_____
_____
_____
_____
_____
_____
_____
_____
_____
_____
_____
_____
_____
_____
_____
_____
_____
_____
_____
_____
_____
_____
_____
_____
_____
_____

FIFRA GLP INSPECTION CHECKLIST
Data Audit Review
Revised 9/93

Laboratory: _____ Insp. Init.: _____ Date: _____

FORM I — GLP COMPLIANCE REVIEW (Continued)

SUBPART B — ORGANIZATION & PERSONNEL			YES	NO	N/A	REMARKS
§160.35(b)	(3)	Perform periodic QA inspections and maintain proper records of each inspection?				
		- What aspects of the ongoing study have been inspected to this point? When?				
	(4)	Periodically submit to management and study director written status reports on each study, noting any problems and corrective actions taken?				
	(5)	Keep dates indicating when management and the study director were notified of inspection findings?				
	(6)	Determine that no deviations were made without proper authorization and documentation?				
	(c)	Are the responsibilities and procedures, records, and indexing methods recorded in writing?				
	(d)	Were these procedures available for review?				

SUBPART C — FACILITIES			YES	NO	N/A	REMARKS
§160.41	General					
	Is the facility's physical layout appropriate to the study?					
	Is there an appropriate degree of separation between/among testing facilities to ensure an appropriate study environment?					
§160.43	Testing system care facilities					
	Do the test system care facilities have:					
	(a) Sufficient number of animal rooms for proper separation of species and projects?					
	(1) Are plants or aquatic animals housed in separate chambers or aquaria?					
	(2) Are aquatic toxicity tests isolated for individual projects?					
	(b) Sufficient number of areas to ensure isolation of studies involving biohazardous substances, including volatile substances, aerosols, radioactive materials, and infectious agents?					
	(c) Separate, isolated areas provided for the diagnosis, treatment, and control of laboratory test system diseases?					
	(d) Proper provisions for handling the collection and disposal of contaminated water, soil, other spent materials, or animal waste handled in order to minimize vermin infestation, odors, disease hazards, and environmental contamination?					
	(e) Provisions to regulate environmental conditions (e.g., temperature, humidity, photoperiod) as specified in the protocol?					

FIFRA GLP INSPECTION CHECKLIST
Compliance Review
Revised 9/93

Laboratory: _____ Insp. Init.: _____ Date: _____

Comments (Please refer to subpart, section, or page numbers):

_____
_____
_____
_____
_____
_____
_____
_____
_____
_____
_____
_____
_____
_____
_____
_____
_____
_____
_____
_____
_____
_____
_____
_____
_____
_____
_____
_____
_____
_____
_____

FIFRA GLP INSPECTION CHECKLIST
Data Audit Review
Revised 9/93

Laboratory: _____ Insp. Init.: _____ Date: _____

FORM I — GLP COMPLIANCE REVIEW (Continued)

SUBPART C — FACILITIES			YES	NO	N/A	REMARKS
§160.43	(f)	<u>For marine organisms:</u> is there an adequate supply of clean seawater as specified in the protocol?				
	(g)	<u>For fresh water organisms:</u> Is there an adequate supply of clean water as specified in the protocol?				
	(h)	<u>For plants:</u> Is there an adequate supply of soil as specified in the protocol?				
§160.45	**Test system supply facilities**					
	Do the test system supply facilities have:					
	(a) Storage areas for feed nutrients, soils, and bedding separate from areas where the test systems are located and protected against infestation and contamination?					
	- Appropriate means for preservation of perishable supplies?					
	(b) The following plant facilities, as specified in the protocol?					
	(1) Facilities for holding, culturing, and maintaining algae and aquatic plants?					
	(2) Facilities for plant growth (e.g., greenhouses, growth chambers, light banks, and fields)?					
	(c) Aquatic animal test facilities, including aquaria, holding tanks, ponds, and ancillary equipment, as specified in the protocol?					
§160.47	**Facilities for handling test, control, and reference substances**					
	Are separate areas for handling test, control, and reference substances provided, including:					
	(a) To prevent contamination or mixups:					
	(1) Separate areas for receipt and storage of substances?					
	(2) Separate areas for mixing substances with a carrier?					
	(3) Separate storage areas for mixtures?					
	- Are these areas separate from those housing the test systems?					
§160.49	**Laboratory operation areas**					
	Is separate laboratory space provided to perform routine and specialized procedures as required by studies?					
§160.51	**Specimen and data storage facilities**					
	Is space provided for archives?					
	Is access to the archives limited?					

FIFRA GLP INSPECTION CHECKLIST
Compliance Review
Revised 9/93

Laboratory: _____ Insp. Init.: _____ Date: _____

Comments (Please refer to subpart, section, or page numbers):

_____
_____
_____
_____
_____
_____
_____
_____
_____
_____
_____
_____
_____
_____
_____
_____
_____
_____
_____
_____
_____
_____
_____
_____
_____
_____
_____
_____
_____
_____
_____

FIFRA GLP INSPECTION CHECKLIST
Data Audit Review
Revised 9/93

Laboratory: _____ Insp. Init.: _____ Date: _____

FORM I — GLP COMPLIANCE REVIEW (Continued)

SUBPART D — EQUIPMENT	YES	NO	N/A	REMARKS
§160.61 **Equipment design**				
Is equipment used in the generation of data and facility environmental control of appropriate design and adequate capacity to function according to protocol requirements?				
Is the equipment in a suitable location for operation, inspection, cleaning, and maintenance?				
§160.63 **Maintenance and calibration of equipment**				
(a) Was equipment adequately inspected, maintained, and calibrated/standardized as required?				
(b) Do the SOPs adequately address the methods, materials, and schedules to be used in the routine inspection, cleaning, maintenance, testing, and calibration/standardization of equipment, including action taken in case of a malfunction?				
Is a specific contact person responsible for the performance of each operation?				
(c) Are written records maintained of all inspection, maintenance, testing, and/or calibrating/standardization operations?				
- Do these records describe whether the maintenance operations were routine and followed the SOPs?				
- Are written records kept of all **non-routine** repairs performed as a result of failure or malfunction?				
- Do the non-routine records document the nature of the defect, how and when the defect was discovered, and the remedial action taken in response?				
- Are the records signed or initialled and dated by the person making the entries?				

SUBPART E — TESTING FACILITIES OPERATIONS	YES	NO	N/A	REMARKS
§160.81 **Standard Operating Procedures**				
(a) Are written SOPs available and adequate?				
- Are deviations from the SOP adequately documented in the raw data?				
- Are significant changes properly authorized in writing by management?				
(b) Are written SOPs available for the following:				
(1) Test system area preparation?				
(2) Test system care?				

FIFRA GLP INSPECTION CHECKLIST
Compliance Review
Revised 9/93

Laboratory: _____ Insp. Init.: _____ Date: _____

Comments (Please refer to subpart, section, or page numbers):

_____
_____
_____
_____
_____
_____
_____
_____
_____
_____
_____
_____
_____
_____
_____
_____
_____
_____
_____
_____
_____
_____
_____
_____
_____
_____
_____
_____
_____
_____

FIFRA GLP INSPECTION CHECKLIST
Data Audit Review
Revised 9/93

Laboratory: _____ Insp. Init.: _____ Date: _____

FORM I — GLP COMPLIANCE REVIEW (Continued)

SUBPART E — TESTING FACILITIES OPERATIONS			YES	NO	N/A	REMARKS
§160.81(a)	(3)	Receipt, ID, storage, handling, mixing, and method of sampling of the test, control, and reference substances?				
	(4)	Test system observations?				
	(5)	Laboratory or other tests?				
	(6)	Handling of test systems found moribund or dead?				
	(7)	Necropsy of test systems or postmortem examination of test systems?				
	(8)	Collection and ID of specimens?				
	(9)	Histopathology?				
	(10)	Data handling, storage, and retrieval?				
	(11)	Maintenance and calibration of equipment?				
	(12)	Transfer, proper placement, and ID of test systems?				
	(c)	Are the latest revisions of relevant SOPs available to each work area?				
	(d)	Is a historical file of SOPs and dates of revisions maintained?				
§160.83		**Reagents and solutions**				
		Are all reagents and solutions labeled to indicate identity, concentration, storage requirements, and expiration date?				
		- Are all materials within expiration date?				
§160.90		**Animal and other test system care**				
	(a)	Are SOPs available for housing, feeding, handling, and care of test systems?				
	(b)	Are newly received test systems isolated, and their health status and appropriateness evaluated?				
		- Are these evaluations performed with acceptable veterinary or scientific methods?				
	(c)	At the initiation of the study, were test systems free of disease for the study?				
		- If, during the study, a disease or condition developed, were test systems isolated?				
		- Were test systems treated for the condition in such a manner that treatment did not interfere with the study?				
		- Were the diagnosis, authorization of treatment, description of treatment, and dates of treatment documented in the raw data?				

FIFRA GLP INSPECTION CHECKLIST
Compliance Review
Revised 9/93

Laboratory: _____ Insp. Init.: _____ Date: _____

Comments (Please refer to subpart, section, or page numbers):

_____
_____
_____
_____
_____
_____
_____
_____
_____
_____
_____
_____
_____
_____
_____
_____
_____
_____
_____
_____
_____
_____
_____
_____
_____
_____
_____
_____
_____
_____
_____
_____
_____
_____

FIFRA GLP INSPECTION CHECKLIST
Data Audit Review
Revised 9/93

Laboratory: _____ Insp. Init.: _____ Date: _____

FORM I — GLP COMPLIANCE REVIEW (Continued)

SUBPART E — TESTING FACILITIES OPERATIONS	YES	NO	N/A	REMARKS
§160.90 (d) Were test systems needing to be removed from their housing units adequately identified (e.g., tattoo, color code, ear tag, ear punch, etc.)?				
- Were test system housing units adequately identified?				
(e) Were different species housed in separate rooms as necessary?				
- Were test systems of the same species used for different studies housed in separate rooms?				
- If the species were not housed in separate rooms, was adequate differentiation by space and identification made?				
(1) Were plants, invertebrate animals, and aquatic vertebrate animals used in multispecies tests, if housed in the same room, segregated to avoid mix-up or cross contamination?				
(f) Were cages, racks, pens, enclosures, aquaria, holding tanks, ponds, growth chambers, and other holding, rearing, and breeding areas, and accessory equipment cleaned and sanitized at appropriate intervals?				
(g) Were feed, soil, and water analyzed periodically for contaminants?				
- Was documentation maintained for these analyses?				
(h) Was the bedding used of a type that would not interfere with the conduct of the study?				
- Was the bedding changed as often as necessary?				
(i) If any pest control materials were used, was their use documented?				
- Were pest control materials used that would not interfere with the study?				
(j) Were test systems acclimated to the environmental conditions of the test?				

SUBPART F — TEST, CONTROL, AND REFERENCE SUBSTANCES

§160.105 Test, control, and reference substance characterization

(a) Have the substances been characterized?	Test	Control	Reference	Documentation
- Identity				
- Strength				
- Purity				
- Stability				
- Uniformity				

FIFRA GLP INSPECTION CHECKLIST
Compliance Review
Revised 9/93

Laboratory: _____ Insp. Init.: _____ Date: _____

Comments (Please refer to subpart, section, or page numbers):

_____
_____
_____
_____
_____
_____
_____
_____
_____
_____
_____
_____
_____
_____
_____
_____
_____
_____
_____
_____
_____
_____
_____
_____
_____
_____
_____
_____
_____
_____
_____
_____

FIFRA GLP INSPECTION CHECKLIST
Data Audit Review
Revised 9/93

Laboratory: _____ Insp. Init.: _____ Date: _____

FORM I — GLP COMPLIANCE REVIEW (Continued)

SUBPART F — TEST, CONTROL, AND REFERENCE SUBSTANCES	YES	NO	N/A	REMARKS
§160.105 - Were methods of synthesis, fabrication, or derivation of the test, control, or reference substance documented?				
- Was the location of documentation specified?				
(b) Were the solubility and/or stability of the substance determined before the experiment start date?				
(c) Did each storage container for a test, control, or reference substance include the following information:				
- name, chemical abstracts service number (CAS) or code number?				
- batch number?				
- expiration date, if any?				
- storage conditions, if appropriate?				
- Were storage containers assigned to a particular test substance for the duration of the study?				
(d) For studies of more than 4 weeks experimental duration, were reserve samples from each batch of test, control, and reference substances retained for the period of time provided in §160.195?				
- Where are reserve samples archived?				
(e) Was the stability of the substance under the storage conditions at the test site known for all studies?				
§160.107 Test, control, and reference substance handling				
(a) Did an SOP covering handling of substances exist?				
(b) Were the substances stored according to the SOP?				
(c) Was distribution made so as to preclude the possibility of contamination, deterioration, or damage?				
(d) Was proper ID of substances maintained throughout the distribution process?				
(e) Was documentation maintained, including date and quantity of each receipt and distribution?				
§160.113 Mixtures of substances with carriers				
(a) Was appropriate analytical testing performed for each test, control, or reference substance:				
(1) To determine uniformity?				
- To determine, periodically, the concentration of the test, control, or reference substance in the mixture?				
(2) To determine solubility in the mixture, if necessary?				
- Was solubility testing done before the experimental start date?				

FIFRA GLP INSPECTION CHECKLIST
Compliance Review
Revised 9/93

Laboratory: _____ Insp. Init.: _____ Date: _____

Comments (Please refer to subpart, section, or page numbers):

_____
_____
_____
_____
_____
_____
_____
_____
_____
_____
_____
_____
_____
_____
_____
_____
_____
_____
_____
_____
_____
_____
_____
_____
_____
_____
_____
_____
_____
_____
_____

Laboratory: _____ Insp. Init.: _____ Date: _____

FORM I — GLP COMPLIANCE REVIEW (Continued)

SUBPART F — TEST, CONTROL, AND REFERENCE SUBSTANCES	YES	NO	N/A	REMARKS
§160.113 (3) To determine the stability in the mixture before the experimental start date or according to the SOP?				
(b) Was the expiration date shown on the mixture container, if necessary?				
(c) Was assurance made that the vehicle did not interfere with the integrity of the test?				

SUBPART G — PROTOCOL FOR AND CONDUCT OF A STUDY	YES	NO	N/A	REMARKS
§160.120 Protocol				
(d) Does the study have an approved written protocol indicating objectives and all methods?				
- Does the protocol contain at least the following:				
(1) A descriptive title and statement of purpose?				
(2) Identification of the test, control, and reference substance by name, CAS number, or code number?				
(3) Name and address of both sponsor and testing facility?				
(4) Proposed experimental start and termination dates?				
(5) Justification for selection of the test system?				
(6) Where applicable, the number, body weight, range, sex, source of supply, species, strain, substrain, and age of the test system?				
(7) Procedure for identification of the test system?				
(8) Description of the experimental design, including methods for the control of bias?				
(9) A description and/or identification of the:				
- diet used in the study?				
- solvents, emulsifiers and/or other materials used to solubilize or suspend the test, control, or reference substance before mixing with the carrier?				
- Specifications for acceptable levels of contaminants?				

FIFRA GLP INSPECTION CHECKLIST
Compliance Review
Revised 9/93

23

Laboratory: _____ Insp. Init.: _____ Date: _____

Comments (Please refer to subpart, section, or page numbers):

_____
_____
_____
_____
_____
_____
_____
_____
_____
_____
_____
_____
_____
_____
_____
_____
_____
_____
_____
_____
_____
_____
_____
_____
_____
_____
_____
_____
_____

FIFRA GLP INSPECTION CHECKLIST
Data Audit Review
Revised 9/93

Laboratory: _____ Insp. Init.: _____ Date: _____

FORM I — GLP COMPLIANCE REVIEW (Continued)

SUBPART G — PROTOCOL FOR AND CONDUCT OF A STUDY			YES	NO	N/A	REMARKS
§160.120(a)	(10)	Route of administration and reason for its choice?				
	(11)	Dosage level in appropriate units and method and frequency of administration?				
	(12)	Type and frequency of tests, analyses, and measurements to be made?				
	(13)	The records to be maintained?				
	(14)	The date of approval of the protocol by the sponsor?				
		- The dated signature of the study director?				
	(15)	A statement of the proposed statistical method to be used?				
	(e)	Are all changes or revisions and reasons:				
		- documented?				
		- signed by the study director?				
		- dated?				
		- maintained with the protocol?				
§160.130		Conduct of a study				
	(a)	Was the study conducted in accordance with the protocol?				
	(b)	Were the test systems monitored in conformity with the protocol?				
	(c)	Are specimens identified by:				
		- test system?				
		- study?				
		- nature of collection?				
		- date of collection?				
		- Is the specimen information either on the container or accompanying the specimen described in a manner that precludes error?				
	(d)	If applicable, are gross necropsy observations available to the pathologist for the histopathological exam?				
	(e)	Were all data recorded promptly and legibly in ink?				
		- Were all data entries (non-automated) signed (or initialed) and dated on the day of entry?				

FIFRA GLP INSPECTION CHECKLIST
Compliance Review
Revised 9/93

Laboratory: _____ Insp. Init.: _____ Date: _____

Comments (Please refer to subpart, section, or page numbers):

_____
_____
_____
_____
_____
_____
_____
_____
_____
_____
_____
_____
_____
_____
_____
_____
_____
_____
_____
_____
_____
_____
_____
_____
_____
_____
_____
_____
_____
_____
_____
_____
_____
_____
_____

FIFRA GLP INSPECTION CHECKLIST
Data Audit Review
Revised 9/93

Laboratory: _____ Insp. Init.: _____ Date: _____

FORM I — GLP COMPLIANCE REVIEW (Continued)

SUBPART G — PROTOCOL FOR AND CONDUCT OF A STUDY	YES	NO	N/A	REMARKS
§160.30(F) - Were changes in entries made so as not to obscure the original entry?				
- Were reasons given for changes?				
- Were changes identified and dated?				
- For automated data, was the individual responsible for direct data input identified at the time of data input?				
§160.135 **Physical and chemical characterizations studies**				
(a) Were all provisions of the GLP standards applied to physical and chemical characterization studies designed to determine stability, solubility, octanol water partition coefficient, volatility, and persistence of test, control, or reference substances?				
§160.190 **Storage and retrieval of records and data**				
(b) Do archives exist for orderly storage and expedient retrieval of all raw data, documentation, protocols, specimens, and interim and final reports?				
- Are the conditions of the storage area appropriate to minimize deterioration in accordance with the time period of their retention and the nature of the documents or specimens?				
(c) Is an individual responsible for the archives?				
(d) Is it specified that only authorized personnel have access to the archives?				
(e) Is the material retained in the archives indexed for rapid retrieval?				

FIFRA GLP INSPECTION CHECKLIST
Compliance Review
Revised 9/93

Laboratory: _____ Insp. Init.: _____ Date: _____

Comments (Please refer to subpart, section, or page numbers):

_____
_____
_____
_____
_____
_____
_____
_____
_____
_____
_____
_____
_____
_____
_____
_____
_____
_____
_____
_____
_____
_____
_____
_____
_____
_____
_____
_____
_____
_____

FIFRA GLP INSPECTION CHECKLIST
Data Audit Review
Revised 9/93

Laboratory: _____   Insp. Init.: _____   Date: _____

PART III — GLP DATA AUDIT REVIEW

FORM II — DATA AUDIT REVIEW

Please complete this form for each data audit selected.

Study selected for review:

    Test substance: _____

    Study title: _____

    Lab ID No.: _____

    Sponsor (name and address): _____

    Study director: _____

    Study initiation date: _____

    Study completion date: _____

    Aspect of the study audited: _____

SUBPART A — GENERAL PROVISIONS		YES	NO	N/A	REMARKS
§160.10	**Applicability to study performed under grant and contract** Was the laboratory, contractor, or grantee informed that their services must be conducted in compliance with 40 CFR Part 160?				
§160.12	**Compliance statement**				
	(a) Was a compliance statement signed by the applicant?				
	the sponsor?				
	the study director?				
	(b) Was the compliance statement completed and submitted with the study report?				
	(c) Did the compliance statement include any statement of differences from the GLP regulations?				

FIFRA GLP INSPECTION CHECKLIST
Data Audit Review
Revised 9/93

Laboratory: _____ Insp. Init.: _____ Date: _____

Comments (Please refer to subpart, section, or page numbers):

_____
_____
_____
_____
_____
_____
_____
_____
_____
_____
_____
_____
_____
_____
_____
_____
_____
_____
_____
_____
_____
_____
_____
_____
_____
_____
_____
_____
_____
_____
_____
_____
_____

FIFRA GLP INSPECTION CHECKLIST
Data Audit Review
Revised 9/93

Laboratory: _____ Insp. Init.: _____ Date: _____

FORM II — DATA AUDIT REVIEW (Continued)

SUBPART B — ORGANIZATION & PERSONNEL	YES	NO	N/A	REMARKS
§160.29 **Personnel**				
(a) Were training, education, and experience adequate?				
(b) Were training and experience records available?				
(c) Was the number of personnel adequate?				
§160.31 **Testing facility management**				
(a) Was a study director designated prior to study initiation?				
(b) Was the study director replaced during the course of the study?				
If so, was this done promptly?				
(c) Was a quality assurance unit in place?				
(d) Are personnel, resources, facilities, equipment, materials, and methodologies available for inspection?				
(e) Were deviations in the study communicated to the study director and corrective actions taken and documented?				
§160.33 **Study director**				
Did the study director have adequate education, training, and experience?				
Did the study director understand that his/her responsibilities included the following assurances:				
(a) The protocol, including any change, was approved and followed?				
(b) All experimental data were accurately recorded and verified?				
(c) Unforeseen circumstances were noted and corrective action taken and documented?				
(d) Test systems were as specified in the protocol?				
(e) All GLPs were followed?				
(f) All required data was transferred to the archives?				
§160.35 **Quality Assurance Unit**				
(a) Was a separate and independent QAU in place at the time of the study?				

FIFRA GLP INSPECTION CHECKLIST
Data Audit Review
Revised 9/93

Laboratory: _____ Insp. Init.: _____ Date: _____

Comments (Please refer to subpart, section, or page numbers):

_____
_____
_____
_____
_____
_____
_____
_____
_____
_____
_____
_____
_____
_____
_____
_____
_____
_____
_____
_____
_____
_____
_____
_____
_____
_____
_____
_____
_____
_____
_____
_____

FIFRA GLP INSPECTION CHECKLIST
Data Audit Review
Revised 9/93

Laboratory: _____ Insp. Init.: _____ Date: _____

FORM II — DATA AUDIT REVIEW (Continued)

SUBPART B — ORGANIZATION & PERSONNEL		YES	NO	N/A	REMARKS
§160.35	(b) Did the QAU:				
	(1) Maintain a complete copy of the master schedule indexed by test substance? (The required elements include the test substance, test system, nature of study, date initiated, current status, identity of sponsor, name of study director.)				
	(2) Maintain copies of protocols?				
	(3) Perform periodic QA inspections and maintain proper records of each inspection?				
	(4) Periodically submit to management and study director written status reports on each study, noting any problems and corrective actions taken?				
	(5) Keep dates indicating when management and the study director were notified of inspection findings?				
	(6) Determine that no deviations were made without proper authorization and documentation?				
	(c) Are the responsibilities and procedures, records, and indexing methods recorded in writing?				
	(d) Were these procedures available for review?				

SUBPART C — FACILITIES		YES	NO	N/A	REMARKS
§160.51	Specimen and data storage facilities				
	Is space provided for archives?				
	Is access to the archives limited?				

SUBPART D — EQUIPMENT		YES	NO	N/A	REMARKS
§160.61	Equipment design				
	Was the equipment used in the generation of data and facility environmental control appropriately designed and of adequate capacity to function according to protocol requirements?				
§160.63	Maintenance and calibration of equipment				
	(a) Was equipment adequately inspected, maintained, and calibrated/standardized as required?				
	(b) Did the SOPs adequately address the methods, materials, and schedules to be used in routine inspection, cleaning, maintenance, testing, and calibration/standardization of equipment, including action taken in case of a malfunction?				
	(c) Were written records maintained of all inspection, maintenance, testing, and/or calibrating/standardization operations?				

FIFRA GLP INSPECTION CHECKLIST
Data Audit Review
Revised 9/93

Laboratory: _____   Insp. Init.: _____   Date: _____

Comments (Please refer to subpart, section, or page numbers):

_____
_____
_____
_____
_____
_____
_____
_____
_____
_____
_____
_____
_____
_____
_____
_____
_____
_____
_____
_____
_____
_____
_____
_____
_____
_____
_____
_____
_____
_____

FIFRA GLP INSPECTION CHECKLIST
Data Audit Review
Revised 9/93

Laboratory: _____ Insp. Init.: _____ Date: _____

FORM II — DATA AUDIT REVIEW (Continued)

SUBPART D — EQUIPMENT	YES	NO	N/A	REMARKS
§160.63(c) - Did these records describe whether the maintenance operations were routine and followed the SOPs?				
- Were written records kept of all **non-routine** repairs performed as a result of failure or malfunction?				
- Did the non-routine records document the nature of the defect, how and when the defect was discovered, and the remedial action taken in response?				
- Were the records signed or initialled and dated by the person making the entries?				

SUBPART E — TESTING FACILITIES OPERATIONS	YES	NO	N/A	REMARKS
§160.81 Standard Operating Procedures				
(a) Were written SOPs in place during the study adequate and available for review?				
- Are deviations from the SOP adequately documented in the raw data?				
- Were all significant changes properly authorized in writing by management?				
(b) Were written SOPs available for the following:				
(1) Test system area preparation?				
(2) Test system care?				
(3) Receipt, ID, storage, handling, mixing, and method of sampling of the test, control, and reference substances?				
(4) Test system observations?				
(5) Laboratory or other tests?				
(6) Handling of test systems found moribund or dead?				
(7) Necropsy of test systems or postmortem examination of test systems?				
(8) Collection and ID of specimens?				
(9) Histopathology?				
(10) Data handling, storage, and retrieval?				
(11) Maintenance and calibration of equipment?				
(12) Transfer, proper placement, and ID of test systems?				
(c) Is a historical file of SOPs and dates of revisions maintained?				

FIFRA GLP INSPECTION CHECKLIST
Data Audit Review
Revised 9/93

Laboratory: _____ Insp. Init.: _____ Date: _____

Comments (Please refer to subpart, section, or page numbers):

_____
_____
_____
_____
_____
_____
_____
_____
_____
_____
_____
_____
_____
_____
_____
_____
_____
_____
_____
_____
_____
_____
_____
_____
_____
_____
_____
_____
_____
_____
_____
_____

FIFRA GLP INSPECTION CHECKLIST
Data Audit Review
Revised 9/93

Laboratory: _____ Insp. Init.: _____ Date: _____

FORM II — DATA AUDIT REVIEW (Continued)

SUBPART E — TESTING FACILITIES OPERATIONS	YES	NO	N/A	REMARKS
§160.83 Reagents and solutions				
Are records for reagents and solutions available that would indicate identity, concentration, storage requirements, and expiration date?				
§160.90 Animal and other test system care				
(a) Were SOPs for housing, feeding, handling, and care of test systems available?				
(b) At the initiation of the study, were test systems free of disease and appropriate for the study?				
- If test systems developed a disease or condition during the study, were test systems isolated?				
- Were test systems treated for the condition in such a manner that treatment did not interfere with the study?				
- Were the diagnosis, authorization of treatment, description of treatment, and dates of treatment documented in the raw data?				
(c) Were test systems needing to be removed from their housing units adequately identified (e.g., tattoo, color code, ear tag, ear punch, etc.)?				
- Were test system housing units adequately identified?				
(d) Were different species housed in separate rooms as necessary?				
- Were test systems of the same species, used for different studies, housed in separate rooms?				
- If the species were not housed in separate rooms, was adequate differentiation by space and identification made?				
(1) Were plants, invertebrate animals, and aquatic vertebrate animals used in multispecies tests, if housed in the same room, segregated to avoid mix-up or cross contamination?				
(e) Are records available indicating whether cages, racks, pens, enclosures, aquaria, holding tanks, ponds, growth chambers, and other holding, rearing, and breeding areas, and accessory equipment were cleaned and sanitized at appropriate intervals?				
(f) Were feed, soil, and water analyzed periodically for contaminants?				
- Was documentation maintained for these analyses?				
(g) Was the bedding used of a type that would not interfere with the conduct of the study?				
- Was the bedding changed as often as necessary?				

FIFRA GLP INSPECTION CHECKLIST
Data Audit Review
Revised 9/93

Laboratory: _____ Insp. Init.: _____ Date: _____

Comments (Please refer to subpart, section, or page numbers):

_____
_____
_____
_____
_____
_____
_____
_____
_____
_____
_____
_____
_____
_____
_____
_____
_____
_____
_____
_____
_____
_____
_____
_____
_____
_____
_____
_____
_____
_____
_____
_____
_____
_____

FIFRA GLP INSPECTION CHECKLIST
Data Audit Review
Revised 9/93

Laboratory: _____ Insp. Init.: _____ Date: _____

FORM II — DATA AUDIT REVIEW (Continued)

SUBPART E — TESTING FACILITIES OPERATIONS	YES	NO	N/A	REMARKS
§160.90 (h) If any pest control materials were used, was their use documented?				
- Were pest control materials used that would not interfere with the study?				
(i) Were test systems acclimated to the environmental conditions of the test?				

SUBPART F — TEST, CONTROL, AND REFERENCE SUBSTANCES

§160.105 Test, control, and reference substance characterization

(a) Were the substances characterized?	Test	Control	Reference	Documentation
- Identity				
- Strength				
- Purity				
- Stability				
- Uniformity				

SUBPART F — TEST, CONTROL, AND REFERENCE SUBSTANCES	YES	NO	N/A	REMARKS
§160.105(a) - Were methods of synthesis, fabrication, or derivation of the test, control, or reference substance documented and the location specified?				
- Was the location of documentation specified?				
(b) Were the solubility and/or stability of the test substance determined before the experiment start date?				
(c) Did each storage container for a test, control, or reference substance include the following information:				
- name, chemical abstracts service number (CAS) or code number?				
- batch number?				
- expiration date, if any?				
- storage conditions, if appropriate?				
- Were storage containers assigned to a particular test substance for the duration of the study?				
(d) For studies of more than 4 weeks experimental duration, were reserve samples from each batch of test, control, and reference substances retained for the period of time provided in §160.195?				
- Where are reserve samples archived?				
(e) Was the stability of the substance under the storage conditions at the test site known for all studies?				

FIFRA GLP INSPECTION CHECKLIST
Data Audit Review
Revised 9/93

Laboratory: _____   Insp. Init.: _____   Date: _____

Comments (Please refer to subpart, section, or page numbers):

_____
_____
_____
_____
_____
_____
_____
_____
_____
_____
_____
_____
_____
_____
_____
_____
_____
_____
_____
_____
_____
_____
_____
_____
_____
_____
_____
_____
_____
_____
_____
_____
_____
_____

FIFRA GLP INSPECTION CHECKLIST
Data Audit Review
Revised 9/93

Laboratory: _____ Insp. Init.: _____ Date: _____

FORM II — DATA AUDIT REVIEW (Continued)

SUBPART F — TEST, CONTROL, AND REFERENCE SUBSTANCES	YES	NO	N/A	REMARKS
§160.107 Test, control, and reference substance handling				
(a) Did an SOP covering handling of substances exist?				
(b) Were the substances stored according to the SOP?				
(c) Was distribution made so as to preclude the possibility of contamination, deterioration, or damage?				
(d) Was proper ID of substances maintained throughout the distribution process?				
(e) Was documentation maintained, including date and quantity of each receipt and distribution?				
§160.113 Mixtures of substances with carriers				
(a) Was appropriate analytical testing performed for each test, control, or reference substance:				
(1) To determine uniformity?				
- To determine, periodically, the concentration of the test, control, or reference substance in the mixture?				
(2) To determine solubility in the mixture, if necessary?				
- Was solubility testing done before the experimental start date?				
(3) To determine the stability in the mixture before the experimental start date or according to the SOP?				
(b) Was the expiration date shown on the mixture container, if necessary?				
(c) Was assurance made that the vehicle did not interfere with the integrity of the test?				

SUBPART G — PROTOCOL FOR AND CONDUCT OF A STUDY	YES	NO	N/A	REMARKS
§160.120 Protocol				
(d) Did the study have an approved written protocol indicating objectives and all methods?				
- Did each protocol contain at least the following:				
(1) A descriptive title and statement of purpose?				
(2) Identification of the test, control, and reference substance by name, CAS number, or code number?				
(3) Name and address of both sponsor and testing facility?				

FIFRA GLP INSPECTION CHECKLIST
Data Audit Review
Revised 9/93

Laboratory: _____ Insp. Init.: _____ Date: _____

Comments (Please refer to subpart, section, or page numbers):

_____
_____
_____
_____
_____
_____
_____
_____
_____
_____
_____
_____
_____
_____
_____
_____
_____
_____
_____
_____
_____
_____
_____
_____
_____
_____
_____
_____
_____
_____
_____
_____
_____
_____

FIFRA GLP INSPECTION CHECKLIST
Data Audit Review
Revised 9/93

Laboratory: _____ Insp. Init.: _____ Date: _____

FORM II — DATA AUDIT REVIEW (Continued)

SUBPART G — PROTOCOL FOR AND CONDUCT OF A STUDY			YES	NO	N/A	REMARKS
§160.120(a)	(4)	Proposed experimental start and termination dates?				
	(5)	Justification for selection of the test system?				
	(6)	Where applicable, the number, body weight, range, sex, source of supply, species, strain, substrain, and age of the test system?				
	(7)	Procedure for identification of the test system?				
	(8)	Description of the experimental design, including methods for the control of bias?				
	(9)	A description and/or identification of the:				
		- diet used in the study?				
		- solvents, emulsifiers and/or other materials used to solubilize or suspend the test, control, or reference substance before mixing with the carrier?				
		- specifications for acceptable levels of contaminants?				
	(10)	Route of administration and reason for its choice?				
	(11)	Dosage level in appropriate units and method and frequency of administration?				
	(12)	Type and frequency of tests, analyses, and measurements to be made?				
	(13)	The records to be maintained?				
	(14)	The date of approval of the protocol by the sponsor?				
		- The dated signature of the study director?				
	(15)	A statement of the proposed statistical method to be used?				
	(e)	Were all changes or revisions and reasons:				
		- documented?				
		- signed by the study director?				
		- dated?				
		- maintained with the protocol?				
§160.130		Conduct of a study				
	(a)	Was the study conducted in accordance with the protocol?				

FIFRA GLP INSPECTION CHECKLIST
Data Audit Review
Revised 9/93

Laboratory: _____ Insp. Init.: _____ Date: _____

Comments (Please refer to subpart, section, or page numbers):

_____
_____
_____
_____
_____
_____
_____
_____
_____
_____
_____
_____
_____
_____
_____
_____
_____
_____
_____
_____
_____
_____
_____
_____
_____
_____
_____
_____
_____
_____
_____

FIFRA GLP INSPECTION CHECKLIST
Data Audit Review
Revised 9/93

Laboratory: _____ Insp. Init.: _____ Date: _____

FORM II — DATA AUDIT REVIEW (Continued)

SUBPART G — PROTOCOL FOR AND CONDUCT OF A STUDY	YES	NO	N/A	REMARKS
§160.130 (b) Were the test systems monitored in conformity with the protocol?				
(c) Were the specimens identified by:				
- test system?				
- study?				
- nature of collection?				
- date of collection?				
- Was the specimen information either on the container or accompanying the specimen described in a manner that precludes error?				
(d) If applicable, were the gross necropsy observations available to the pathologist for the histopathological exam?				
(e) Were all data recorded promptly and legibly in ink?				
- Were all data entries (non-automated) signed (or initialed) and dated on the day of entry?				
- Were changes in entries made so as not to obscure the original entry?				
- Were reasons given for changes?				
- Were changes identified and dated?				
- For automated data, was the individual responsible for direct data input identified at the time of data input?				
§160.135 Physical and chemical characterizations studies				
(a) Were all provisions of the GLP standards applied to physical and chemical characterization studies designed to determine stability, solubility, octanol water partition coefficient, volatility, and persistence of test, control, or reference substances?				

SUBPART J — RECORDS AND REPORTS	YES	NO	N/A	REMARKS
§160.185 Reporting of study results				
(a) Was a final report prepared to contain at least the following?				
(1) Name and address of the facility performing the study?				
- the dates on which the study was initiated, completed, terminated, or discontinued?				

FIFRA GLP INSPECTION CHECKLIST
Data Audit Review
Revised 9/93

Laboratory: _____ Insp. Init.: _____ Date: _____

Comments (Please refer to subpart, section, or page numbers):

_____
_____
_____
_____
_____
_____
_____
_____
_____
_____
_____
_____
_____
_____
_____
_____
_____
_____
_____
_____
_____
_____
_____
_____
_____
_____
_____
_____
_____
_____
_____
_____

FIFRA GLP INSPECTION CHECKLIST
Data Audit Review
Revised 9/93

Laboratory: _____ Insp. Init.: _____ Date: _____

FORM II — DATA AUDIT REVIEW (Continued)

SUBPART J — RECORDS AND REPORTS			YES	NO	N/A	REMARKS
§160.185(a)	(2)	The objectives and procedures as stated in the approved protocol?				
	(3)	The statistical methods employed?				
	(4)	The test, control, and reference substance identified by name, CAS number or code number, strength, purity, and composition?				
	(5)	Stability and, if needed, solubility, of the substances under conditions of administration?				
	(6)	A description of the methods used?				
	(7)	A description of the test system used?				
		- Where applicable, the number of animals used, sex, body weight range, source of supply, species, strain and substrain, age, and procedures used for ID?				
	(8)	A description of the dosage, dosage regimen, route of administration, and duration?				
	(9)	A description of all of the circumstances that may have affected the quality or integrity of the data?				
	(10)	The name of the study director?				
		- The names of other scientists, professionals, and supervisory personnel?				
	(11)	A description of the transformations, calculations, or operations performed on the data?				
		- A summary and analysis of the data?				
		- A statement of conclusions drawn from the data?				
	(12)	Signed and dated reports of each of the individual scientists or other professionals involved in the study, including each person who conducted an analysis or evaluation of data or specimens?				
	(13)	The locations where all specimens, raw data, and the final report are to be stored?				
	(14)	A QAU statement prepared and signed as specified in §160.35(b)(7)?				
	(b)	Was the final report signed and dated by the study director?				
	(c)	Were corrections or additions to the final report in the form of an amendment by the study director?				

FIFRA GLP INSPECTION CHECKLIST
Data Audit Review
Revised 9/93

Laboratory: _____   Insp. Init.: _____   Date: _____

Comments (Please refer to subpart, section, or page numbers):

FIFRA GLP INSPECTION CHECKLIST
Data Audit Review
Revised 9/93

Laboratory: _____ Insp. Init.: _____ Date: _____

FORM II — DATA AUDIT REVIEW (Continued)

SUBPART J — RECORDS AND REPORTS	YES	NO	N/A	REMARKS
§160.185(c) - Were the amendments clearly identified with:				
- reasons for change?				
- date?				
- signature of person responsible?				
(d) Is a copy of the final report with amendments maintained by the sponsor and the test facility?				
§160.190 Storage and retrieval of records and data				
(a) Where are the raw data for the study archived in compliance with this section?				
(b) Were all raw data, documentation, records, protocols, specimens, and final reports retained which were generated as a result of a study?				
- Were all correspondence and other documents relating to interpretation and evaluation of data, other than those contained in the final report, retained?				
(c) Are archives provided for orderly storage and expedient retrieval of all raw data, documentation, protocols, specimens, and interim and final reports?				
- Are the conditions of the storage area appropriate to minimize deterioration in accordance with the time period of their retention and the nature of the documents or specimens?				
(d) Is an individual responsible for the archives?				
(e) Is it specified that only authorized personnel have access to the archives?				
(f) Is the material retained in the archives indexed for rapid retrieval?				
§160.195 Retention of records				
(1) Does the sponsor hold a research or marketing permit for the test substance?				
If Yes, are the data retained?				
(2) Has the sponsor applied for, but not received, a research or marketing permit?				
If Yes, were the data retained for at least five years?				
(3) If the answer to (a) or (b) above is No, were the data retained for at least two years?				
(d) Are wet specimens, samples of test, control, or reference substances, and specially prepared material that are relatively fragile and differ markedly in stability and quality during storage retained only as long as the quality of the preparation affords evaluation?				

FIFRA GLP INSPECTION CHECKLIST
Data Audit Review
Revised 9/93

Laboratory: _____ Insp. Init.: _____ Date: _____

Comments (Please refer to subpart, section, or page numbers):

_____
_____
_____
_____
_____
_____
_____
_____
_____
_____
_____
_____
_____
_____
_____
_____
_____
_____
_____
_____
_____
_____
_____
_____
_____
_____
_____
_____
_____
_____
_____
_____
_____
_____
_____

FIFRA GLP INSPECTION CHECKLIST
Data Audit Review
Revised 9/93

Laboratory: _____
Insp. Init.: _____ Date: _____

FORM II — DATA AUDIT REVIEW (Continued)

SUBPART J — RECORDS AND REPORTS	YES	NO	N/A	REMARKS
§160.195 (e) Were the master schedule sheet, copies of protocols, and records of quality assurance inspections, as required by 160.35(c), maintained by the QAU as an easily-accessible system of records for the period of time specified in questions 1 or 2 of this section?				
(f) Were provisions in place to retain summaries of training, experience, and job descriptions, required to be maintained by 160.29(b), as well as all other testing facility employment records for the length of time specified in questions 1 or 2 of this section?				
(g) Were provisions in place to retain records and reports of the maintenance, calibration, and inspection of equipment, as required by 160.63(b) and (c), for the length of time specified in questions 1 or 2 of this section?				
(h) Were provisions in place to retain records required by this part either as original records or as true copies such as photocopies, microfilm, microfiche, or other accurate reproductions of the original records?				
§169.2(k) 40 CFR 169 Books and Records — Does the sponsor currently hold a research or marketing permit for the test substance?				
If Yes, were the *original* raw data for this study retained?				
If Yes, where are raw data retained?				

FIFRA GLP INSPECTION CHECKLIST
Data Audit Review
Revised 9/93

Appendix E

GLP TSCA Compliance Checklist

UNITED STATES ENVIRONMENTAL PROTECTION AGENCY
GLP COMPLIANCE INSPECTION CHECKLIST

PART I — GENERAL

GLP FACILITY INSPECTION

Name: _____ Date: _____

Address: _____ Insp. No. _____

City: _____ State: _____ Zip: _____

Phone No.: _____ Contact Person: _____

FACILITY INFORMATION:

Is this facility:

    a sponsor lab? _____
    a contractor lab? _____
    a management company? _____

What types of studies are conducted here (i.e., toxicology, chemical analysis, field) _____

PREINSPECTION REVIEW: (Obtained from Regional Office Files)

Date(s) of Previous EPA Inspection(s): _____

Previous Findings:
_____
_____
_____

REASON FOR INSPECTION: The purpose of this inspection is to determine if the facility is in compliance with the requirements of TSCA, codified in 40 CFR Part 792.

☐ Randomly Selected Neutral Inspection
☐ Selected for Cause
    ☐ Referral from _____
    ☐ Other: (Specify) _____

TSCA GLP INSPECTION CHECKLIST
General Information
Revised 9/93

Laboratory: _____ Insp. Init.: _____ Date: _____

Comments (Please refer to subpart, section, or page numbers):

TSCA GLP INSPECTION CHECKLIST
Data Audit Review
Revised 9/93

Laboratory: _____ Insp. Init.: _____ Date: _____

OPENING CONFERENCE

PRELIMINARY INFORMATION

1. Laboratory personnel present and interviewed:

   Name: _____ Title: _____
   Name: _____ Title: _____
   Name: _____ Title: _____
   Name: _____ Title: _____
   Name: _____ Title: _____

2. EPA inspector accompanied:

   Name: _____ Agency: _____
   Name: _____ Agency: _____

3. Credentials presented to: _____

4. "Notice of Inspection" signed by laboratory official and copy provided to official? ☐ Yes ☐ No

5. Was a GLP Compliance Review conducted? ☐ Yes ☐ No
   If so, complete Form I.

6. Was a data audit (or audits) conducted? ☐ Yes ☐ No
   If so, complete Form II for each study audited.
   List of studies audited:
   _____
   _____
   _____

CLOSING CONFERENCE (to be completed at conclusion of the inspection)

A. Date: _____ Time: _____ Where conducted: _____
   Facility Representative(s) Present:

   Name: _____ Title: _____
   Name: _____ Title: _____
   Name: _____ Title: _____

TSCA GLP INSPECTION CHECKLIST
General Information
Revised 9/93

Laboratory: _____ Insp. Init.: _____ Date: _____

Comments (Please refer to subpart, section, or page numbers):

_____
_____
_____
_____
_____
_____
_____
_____
_____
_____
_____
_____
_____
_____
_____
_____
_____
_____
_____
_____
_____
_____
_____
_____
_____
_____
_____
_____
_____
_____
_____
_____
_____
_____

TSCA GLP INSPECTION CHECKLIST
Data Audit Review
Revised 9/93

Laboratory: _____ Insp. Init.: _____ Date: _____

CLOSING CONFERENCE (to be completed at conclusion of the inspection)

B. Were facility officials provided copies of:

☐ Receipt for Samples and Documents   ☐ Inspection Confidentiality Notice
☐ Updated Regulations/Guidances   ☐ Declaration of Confidential Business Information

C. Were any documents, records, etc. requested from the facility?   ☐ yes   ☐ no
(If yes, include the list of information requested, and when it is due to be sent)

D. Does the inspector need to conduct any further follow-up activities?   ☐ yes   ☐ no
(If yes, please attach an explanation of what must be done, and a projected schedule for the completion of all follow-up activities.)

Inspector's Signature: _____   Date of Signature: _____

NOTES:

TSCA GLP INSPECTION CHECKLIST
General Information
Revised 9/93

Laboratory: _____ Insp. Init.: _____ Date: _____

Comments (Please refer to subpart, section, or page numbers):

_____
_____
_____
_____
_____
_____
_____
_____
_____
_____
_____
_____
_____
_____
_____
_____
_____
_____
_____
_____
_____
_____
_____
_____
_____
_____
_____
_____
_____
_____
_____
_____

TSCA GLP INSPECTION CHECKLIST
Data Audit Review
Revised 9/93

Laboratory: _____ Insp. Init.: _____ Date: _____

PART II — GLP COMPLIANCE REVIEW CHECKLIST

FORM I — GLP COMPLIANCE REVIEW

Were any ongoing studies available? Please complete this form for <u>each</u> ongoing study selected.

Study selected for review:

Test substance: _____

Study title: _____

Lab ID No.: _____

Sponsor (name and address): _____

Study director: _____

Study initiation date: _____

Proposed completion date: _____

GENERAL INSTRUCTIONS/INFORMATION
1. For any "No" answers, provide explanation.
2. Remarks can be continued in the "Comments" section on the back of each page.
3. Place a line through any item missing. For example, "...name/~~signature~~..."

SUBPART A — GENERAL PROVISIONS	YES	NO	N/A	REMARKS
§792.10 Applicability to studies performed under grants and contracts				
Has laboratory, contractor, or grantee been informed that their services must be conducted in compliance with 40 CFR Part 792?				

SUBPART B — ORGANIZATION & PERSONNEL	YES	NO	N/A	REMARKS
§792.29 Personnel				
(a) Are training, education, and experience adequate?				
(b) Are training and experience records available?				
(c) Is the number of personnel adequate?				
(d) Are personnel health and sanitation precautions being followed?				
(e) Is appropriate clothing available and worn as needed?				
(f) Are any personnel ill to the extent that they have an adverse effect on the study?				
- If so, are they excluded from direct contact with test systems and substances?				

TSCA GLP INSPECTION CHECKLIST
Compliance Review
Revised 9/93

Laboratory: _____ Insp. Init.: _____ Date: _____

Comments (Please refer to subpart, section, or page numbers):

_____
_____
_____
_____
_____
_____
_____
_____
_____
_____
_____
_____
_____
_____
_____
_____
_____
_____
_____
_____
_____
_____
_____
_____
_____
_____
_____
_____
_____
_____

TSCA GLP INSPECTION CHECKLIST
Data Audit Review
Revised 9/93

Laboratory: _____ Insp. Init.: _____ Date: _____

FORM I — GLP COMPLIANCE REVIEW (Continued)

SUBPART B — ORGANIZATION & PERSONNEL	YES	NO	N/A	REMARKS
§792.31 Testing facility management				
(a) Was a study director designated for ongoing study prior to study initiation?				
(b) Has the study director been replaced?				
If so, was this done promptly?				
(c) Is a quality assurance unit in place?				
(d) Are personnel, resources, facilities, equipment, materials, and methodologies available as scheduled?				
(e) Do personnel clearly understand the functions they are to perform?				
(f) Have deviations in the study been communicated to the study director, and have corrective actions been taken and documented?				
§792.33 Study director				
Does the study director have adequate education, training, and experience?				
Is s/he familiar with all aspects of the study?				
Does the study director understand that his/her responsibilities include the following assurances:				
(a) The protocol, including any change, is approved and followed?				
(b) All experimental data are accurately recorded and verified?				
(c) Unforeseen circumstances have been noted and corrective actions taken and documented?				
(d) Test systems are as specified in the protocol?				
(e) All GLPs are followed?				
(f) All data, as required, were transferred to the archives?				
§792.35 Quality Assurance Unit				
(a) Was a separate and independent QAU in place at the time of the study?				
(b) Did the QAU:				
(1) Maintain a complete copy of the master schedule indexed by test substance? (The required elements include the test substance, test system, nature of study, date initiated, current status, identity of sponsor, name of study director.)				
(2) Maintain copies of protocols?				

TSCA GLP INSPECTION CHECKLIST
Compliance Review
Revised 9/93

Laboratory: _____ Insp. Init.: _____ Date: _____

Comments (Please refer to subpart, section, or page numbers):

_____
_____
_____
_____
_____
_____
_____
_____
_____
_____
_____
_____
_____
_____
_____
_____
_____
_____
_____
_____
_____
_____
_____
_____
_____
_____
_____
_____
_____
_____

TSCA GLP INSPECTION CHECKLIST
Data Audit Review
Revised 9/93

Laboratory: _____ Insp. Init.: _____ Date: _____

FORM I — GLP COMPLIANCE REVIEW (Continued)

SUBPART B — ORGANIZATION & PERSONNEL			YES	NO	N/A	REMARKS
§792.35(b)	(3)	Perform periodic QA inspections and maintain proper records of each inspection?				
		- What aspects of the ongoing study have been inspected to this point? When?				
	(4)	Periodically submit to management and study director written status reports on each study, noting any problems and corrective actions taken?				
	(5)	Keep dates indicating when management and the study director were notified of inspection findings?				
	(6)	Determine that no deviations were made without proper authorization and documentation?				
	(c)	Are the responsibilities and procedures, records, and indexing methods recorded in writing?				
	(d)	Were these procedures available for review?				

SUBPART C — FACILITIES			YES	NO	N/A	REMARKS
§792.41	General					
	Is the facility's physical layout appropriate to the study?					
	Is there an appropriate degree of separation between/among testing facilities to ensure an appropriate study environment?					
§792.43	Testing system care facilities					
	Do the test system care facilities have:					
	(a) Sufficient number of animal rooms for proper separation of species and projects?					
	(1) Are plants or aquatic animals housed in separate chambers or aquaria?					
	(2) Are aquatic toxicity tests isolated for individual projects?					
	(b) Sufficient number of areas to ensure isolation of studies involving biohazardous substances, including volatile substances, aerosols, radioactive materials, and infectious agents?					
	(c) Separate, isolated areas provided for the diagnosis, treatment, and control of laboratory test system diseases?					
	(d) Proper provisions for handling the collection and disposal of contaminated water, soil, other spent materials, or animal waste handled in order to minimize vermin infestation, odors, disease hazards, and environmental contamination?					
	(e) Provisions to regulate environmental conditions (e.g., temperature, humidity, photoperiod) as specified in the protocol?					

Laboratory: _____ Insp. Init.: _____ Date: _____

Comments (Please refer to subpart, section, or page numbers):

_____
_____
_____
_____
_____
_____
_____
_____
_____
_____
_____
_____
_____
_____
_____
_____
_____
_____
_____
_____
_____
_____
_____
_____
_____
_____
_____
_____
_____
_____

Laboratory: _____ Insp. Init.: _____ Date: _____

FORM I — GLP COMPLIANCE REVIEW (Continued)

SUBPART C — FACILITIES		YES	NO	N/A	REMARKS
§792.43	(f) <u>For marine organisms:</u> is there an adequate supply of clean seawater as specified in the protocol?				
	(g) <u>For fresh water organisms:</u> Is there an adequate supply of clean water as specified in the protocol?				
	(h) <u>For plants:</u> Is there an adequate supply of soil as specified in the protocol?				
§792.45	Test system supply facilities				
	Do the test system supply facilities have:				
	(a) Storage areas for feed nutrients, soils, and bedding separate from areas where the test systems are located and protected against infestation and contamination?				
	- Appropriate means for preservation of perishable supplies?				
	(b) The following plant facilities, as specified in the protocol?				
	(1) Facilities for holding, culturing, and maintaining algae and aquatic plants?				
	(2) Facilities for plant growth (e.g., greenhouses, growth chambers, light banks, and fields)?				
	(c) Aquatic animal test facilities, including aquaria, holding tanks, ponds, and ancillary equipment, as specified in the protocol?				
§792.47	Facilities for handling test, control, and reference substances				
	Are separate areas for handling test, control, and reference substances provided, including:				
	(a) To prevent contamination or mixups:				
	(1) Separate areas for receipt and storage of substances?				
	(2) Separate areas for mixing substances with a carrier?				
	(3) Separate storage areas for mixtures?				
	- Are these areas separate from those housing the test systems?				
§792.49	Laboratory operation areas				
	Is separate laboratory space provided to perform routine and specialized procedures as required by studies?				
§792.51	Specimen and data storage facilities				
	Is space provided for archives?				
	Is access to the archives limited?				

TSCA GLP INSPECTION CHECKLIST
Compliance Review
Revised 9/93

Laboratory: _____ Insp. Init.: _____ Date: _____

Comments (Please refer to subpart, section, or page numbers):

 TSCA GLP INSPECTION CHECKLIST
Data Audit Review
Revised 9/93

Laboratory: _____ Insp. Init.: _____ Date: _____

FORM I — GLP COMPLIANCE REVIEW (Continued)

SUBPART D — EQUIPMENT	YES	NO	N/A	REMARKS
§792.61 Equipment design				
Is equipment used in the generation of data and facility environmental control of appropriate design and adequate capacity to function according to protocol requirements?				
Is the equipment in a suitable location for operation, inspection, cleaning, and maintenance?				
§792.63 Maintenance and calibration of equipment				
(a) Was equipment adequately inspected, maintained, and calibrated/standardized as required?				
(b) Do the SOPs adequately address the methods, materials, and schedules to be used in the routine inspection, cleaning, maintenance, testing, and calibration/standardization of equipment, including action taken in case of a malfunction?				
Is a specific contact person responsible for the performance of each operation?				
(c) Are written records maintained of all inspection, maintenance, testing, and/or calibrating/standardization operations?				
- Do these records describe whether the maintenance operations were routine and followed the SOPs?				
- Are written records kept of all **non-routine** repairs performed as a result of failure or malfunction?				
- Do the non-routine records document the nature of the defect, how and when the defect was discovered, and the remedial action taken in response?				
- Are the records signed or initialled and dated by the person making the entries?				

SUBPART E — TESTING FACILITIES OPERATIONS	YES	NO	N/A	REMARKS
§792.81 Standard Operating Procedures				
(a) Are written SOPs available and adequate?				
- Are deviations from the SOP adequately documented in the raw data?				
- Are significant changes properly authorized in writing by management?				
(b) Are written SOPs available for the following:				
(1) Test system area preparation?				
(2) Test system care?				

TSCA GLP INSPECTION CHECKLIST
Compliance Review
Revised 9/93

Laboratory: _____ Insp. Init.: _____ Date: _____

Comments (Please refer to subpart, section, or page numbers):

_____
_____
_____
_____
_____
_____
_____
_____
_____
_____
_____
_____
_____
_____
_____
_____
_____
_____
_____
_____
_____
_____
_____
_____
_____
_____
_____
_____
_____
_____
_____
_____
_____
_____
_____
_____

TSCA GLP INSPECTION CHECKLIST
Data Audit Review
Revised 9/93

Laboratory: _____ Insp. Init.: _____ Date: _____

FORM I — GLP COMPLIANCE REVIEW (Continued)

SUBPART E — TESTING FACILITIES OPERATIONS			YES	NO	N/A	REMARKS
§792.81(a)	(3)	Receipt, ID, storage, handling, mixing, and method of sampling of the test, control, and reference substances?				
	(4)	Test system observations?				
	(5)	Laboratory or other tests?				
	(6)	Handling of test systems found moribund or dead?				
	(7)	Necropsy of test systems or postmortem examination of test systems?				
	(8)	Collection and ID of specimens?				
	(9)	Histopathology?				
	(10)	Data handling, storage, and retrieval?				
	(11)	Maintenance and calibration of equipment?				
	(12)	Transfer, proper placement, and ID of test systems?				
	(c)	Are the latest revisions of relevant SOPs available to each work area?				
	(d)	Is a historical file of SOPs and dates of revisions maintained?				
§792.83		**Reagents and solutions**				
		Are all reagents and solutions labeled to indicate identity, concentration, storage requirements, and expiration date?				
		- Are all materials within expiration date?				
§792.90		**Animal and other test system care**				
	(a)	Are SOPs available for housing, feeding, handling, and care of test systems?				
	(b)	Are newly received test systems isolated, and their health status and appropriateness evaluated?				
		- Are these evaluations performed with acceptable veterinary or scientific methods?				
	(c)	At the initiation of the study, were test systems free of disease for the study?				
		- If, during the study, a disease or condition developed, were test systems isolated?				
		- Were test systems treated for the condition in such a manner that treatment did not interfere with the study?				
		- Were the diagnosis, authorization of treatment, description of treatment, and dates of treatment documented in the raw data?				

TSCA GLP INSPECTION CHECKLIST
Compliance Review
Revised 9/93

Laboratory: _____ Insp. Init.: _____ Date: _____

Comments (Please refer to subpart, section, or page numbers):

_____
_____
_____
_____
_____
_____
_____
_____
_____
_____
_____
_____
_____
_____
_____
_____
_____
_____
_____
_____
_____
_____
_____
_____
_____
_____
_____
_____
_____
_____
_____
_____
_____

TSCA GLP INSPECTION CHECKLIST
Data Audit Review
Revised 9/93

Laboratory: _____ Insp. Init.: _____ Date: _____

FORM I — GLP COMPLIANCE REVIEW (Continued)

SUBPART E — TESTING FACILITIES OPERATIONS			YES	NO	N/A	REMARKS
§792.90	(d)	Were test systems needing to be removed from their housing units adequately identified (e.g., tattoo, color code, ear tag, ear punch, etc.)?				
		- Were test system housing units adequately identified?				
	(e)	Were different species housed in separate rooms as necessary?				
		- Were test systems of the same species used for different studies housed in separate rooms?				
		- If the species were not housed in separate rooms, was adequate differentiation by space and identification made?				
	(1)	Were plants, invertebrate animals, and aquatic vertebrate animals used in multispecies tests, if housed in the same room, segregated to avoid mix-up or cross contamination?				
	(f)	Were cages, racks, pens, enclosures, aquaria, holding tanks, ponds, growth chambers, and other holding, rearing, and breeding areas, and accessory equipment cleaned and sanitized at appropriate intervals?				
	(g)	Were feed, soil, and water analyzed periodically for contaminants?				
		- Was documentation maintained for these analyses?				
	(h)	Was the bedding used of a type that would not interfere with the conduct of the study?				
		- Was the bedding changed as often as necessary?				
	(i)	If any pest control materials were used, was their use documented?				
		- Were pest control materials used that would not interfere with the study?				
	(j)	Were test systems acclimated to the environmental conditions of the test?				

SUBPART F — TEST, CONTROL, AND REFERENCE SUBSTANCES

§792.105 Test, control, and reference substance characterization

		Test	Control	Reference	Documentation
(a)	Have the substances been characterized?				
	- Identity				
	- Strength				
	- Purity				
	- Stability				
	- Uniformity				

TSCA GLP INSPECTION CHECKLIST
Compliance Review
Revised 9/93

Laboratory: _____ Insp. Init.: _____ Date: _____

Comments (Please refer to subpart, section, or page numbers):

_____
_____
_____
_____
_____
_____
_____
_____
_____
_____
_____
_____
_____
_____
_____
_____
_____
_____
_____
_____
_____
_____
_____
_____
_____
_____
_____
_____
_____
_____
_____
_____

TSCA GLP INSPECTION CHECKLIST
Data Audit Review
Revised 9/93

Laboratory: _____ Insp. Init.: _____ Date: _____

FORM I — GLP COMPLIANCE REVIEW (Continued)

SUBPART F — TEST, CONTROL, AND REFERENCE SUBSTANCES	YES	NO	N/A	REMARKS
§792.105 - Were methods of synthesis, fabrication, or derivation of the test, control, or reference substance documented?				
- Was the location of documentation specified?				
(b) Were the solubility and/or stability of the substance determined before the experiment start date?				
(c) Did each storage container for a test, control, or reference substance include the following information:				
- name, chemical abstracts service number (CAS) or code number?				
- batch number?				
- expiration date, if any?				
- storage conditions, if appropriate?				
- Were storage containers assigned to a particular test substance for the duration of the study?				
(d) For studies of more than 4 weeks experimental duration, were reserve samples from each batch of test, control, and reference substances retained for the period of time provided in §792.195?				
- Where are reserve samples archived?				
(e) Was the stability of the substance under the storage conditions at the test site known for all studies?				
§792.107 Test, control, and reference substance handling				
(a) Did an SOP covering handling of substances exist?				
(b) Were the substances stored according to the SOP?				
(c) Was distribution made so as to preclude the possibility of contamination, deterioration, or damage?				
(d) Was proper ID of substances maintained throughout the distribution process?				
(e) Was documentation maintained, including date and quantity of each receipt and distribution?				
§792.113 Mixtures of substances with carriers				
(a) Was appropriate analytical testing performed for each test, control, or reference substance:				
(1) To determine uniformity?				
- To determine, periodically, the concentration of the test, control, or reference substance in the mixture?				
(2) To determine solubility in the mixture, if necessary?				
- Was solubility testing done before the experimental start date?				

TSCA GLP INSPECTION CHECKLIST
Compliance Review
Revised 9/93

Laboratory: _____ Insp. Init.: _____ Date: _____

Comments (Please refer to subpart, section, or page numbers):

_____
_____
_____
_____
_____
_____
_____
_____
_____
_____
_____
_____
_____
_____
_____
_____
_____
_____
_____
_____
_____
_____
_____
_____
_____
_____
_____
_____
_____
_____
_____
_____

TSCA GLP INSPECTION CHECKLIST
Data Audit Review
Revised 9/93

Laboratory: _____ Insp. Init.: _____ Date: _____

FORM I — GLP COMPLIANCE REVIEW (Continued)

SUBPART F — TEST, CONTROL, AND REFERENCE SUBSTANCES	YES	NO	N/A	REMARKS
§792.113 (3) To determine the stability in the mixture before the experimental start date or according to the SOP?				
(b) Was the expiration date shown on the mixture container, if necessary?				
(c) Was assurance made that the vehicle did not interfere with the integrity of the test?				

SUBPART G — PROTOCOL FOR AND CONDUCT OF A STUDY	YES	NO	N/A	REMARKS
§792.120 **Protocol**				
(d) Does the study have an approved written protocol indicating objectives and all methods?				
- Does the protocol contain at least the following:				
(1) A descriptive title and statement of purpose?				
(2) Identification of the test, control, and reference substance by name, CAS number, or code number?				
(3) Name and address of both sponsor and testing facility?				
(4) Proposed experimental start and termination dates?				
(5) Justification for selection of the test system?				
(6) Where applicable, the number, body weight, range, sex, source of supply, species, strain, substrain, and age of the test system?				
(7) Procedure for identification of the test system?				
(8) Description of the experimental design, including methods for the control of bias?				
(9) A description and/or identification of the:				
- diet used in the study?				
- solvents, emulsifiers and/or other materials used to solubilize or suspend the test, control, or reference substance before mixing with the carrier?				
- Specifications for acceptable levels of contaminants?				

TSCA GLP INSPECTION CHECKLIST
Compliance Review
Revised 9/93

Laboratory: _____   Insp. Init.: _____   Date: _____

Comments (Please refer to subpart, section, or page numbers):

_____
_____
_____
_____
_____
_____
_____
_____
_____
_____
_____
_____
_____
_____
_____
_____
_____
_____
_____
_____
_____
_____
_____
_____
_____
_____
_____
_____
_____
_____
_____
_____
_____
_____
_____
_____

TSCA GLP INSPECTION CHECKLIST
Data Audit Review
Revised 9/93

Laboratory: _____ Insp. Init.: _____ Date: _____

FORM I — GLP COMPLIANCE REVIEW (Continued)

SUBPART G — PROTOCOL FOR AND CONDUCT OF A STUDY			YES	NO	N/A	REMARKS
§792.120(a)	(10)	Route of administration and reason for its choice?				
	(11)	Dosage level in appropriate units and method and frequency of administration?				
	(12)	Type and frequency of tests, analyses, and measurements to be made?				
	(13)	The records to be maintained?				
	(14)	The date of approval of the protocol by the sponsor?				
		- The dated signature of the study director?				
	(15)	A statement of the proposed statistical method to be used?				
	(e)	Are all changes or revisions and reasons:				
		- documented?				
		- signed by the study director?				
		- dated?				
		- maintained with the protocol?				
§792.130		**Conduct of a study**				
	(a)	Was the study conducted in accordance with the protocol?				
	(b)	Were the test systems monitored in conformity with the protocol?				
	(c)	Are specimens identified by:				
		- test system?				
		- study?				
		- nature of collection?				
		- date of collection?				
		- Is the specimen information either on the container or accompanying the specimen described in a manner that precludes error?				
	(d)	If applicable, are gross necropsy observations available to the pathologist for the histopathological exam?				
	(e)	Were all data recorded promptly and legibly in ink?				
		- Were all data entries (non-automated) signed (or initialed) and dated on the day of entry?				

TSCA GLP INSPECTION CHECKLIST
Compliance Review
Revised 9/93

Laboratory: _____ Insp. Init.: _____ Date: _____

Comments (Please refer to subpart, section, or page numbers):

_____
_____
_____
_____
_____
_____
_____
_____
_____
_____
_____
_____
_____
_____
_____
_____
_____
_____
_____
_____
_____
_____
_____
_____
_____
_____
_____
_____
_____
_____
_____
_____
_____

TSCA GLP INSPECTION CHECKLIST
Data Audit Review
Revised 9/93

Laboratory: _____ Insp. Init.: _____ Date: _____

FORM I — GLP COMPLIANCE REVIEW (Continued)

SUBPART G — PROTOCOL FOR AND CONDUCT OF A STUDY		YES	NO	N/A	REMARKS
§792.30(F)	- Were changes in entries made so as not to obscure the original entry?				
	- Were reasons given for changes?				
	- Were changes identified and dated?				
	- For automated data, was the individual responsible for direct data input identified at the time of data input?				
§792.135	Physical and chemical characterizations studies				
	(a) Were all provisions of the GLP standards applied to physical and chemical characterization studies designed to determine stability, solubility, octanol water partition coefficient, volatility, and persistence of test, control, or reference substances?				
§792.190	Storage and retrieval of records and data				
	(b) Do archives exist for orderly storage and expedient retrieval of all raw data, documentation, protocols, specimens, and interim and final reports?				
	- Are the conditions of the storage area appropriate to minimize deterioration in accordance with the time period of their retention and the nature of the documents or specimens?				
	(c) Is an individual responsible for the archives?				
	(d) Is it specified that only authorized personnel have access to the archives?				
	(e) Is the material retained in the archives indexed for rapid retrieval?				

TSCA GLP INSPECTION CHECKLIST
Compliance Review
Revised 9/93

Laboratory: _____ Insp. Init.: _____ Date: _____

Comments (Please refer to subpart, section, or page numbers):

_____
_____
_____
_____
_____
_____
_____
_____
_____
_____
_____
_____
_____
_____
_____
_____
_____
_____
_____
_____
_____
_____
_____
_____
_____
_____
_____
_____
_____
_____
_____
_____

TSCA GLP INSPECTION CHECKLIST
Data Audit Review
Revised 9/93

Laboratory: _____ Insp. Init.: _____ Date: _____

PART III — GLP DATA AUDIT REVIEW

FORM II — DATA AUDIT REVIEW

Please complete this form for each data audit selected.

Study selected for review:

- Test substance: _____
- Study title: _____
- Lab ID No.: _____
- Sponsor (name and address): _____
- Study director: _____
- Study initiation date: _____
- Study completion date: _____
- Aspect of the study audited: _____

SUBPART A — GENERAL PROVISIONS		YES	NO	N/A	REMARKS
§792.10	**Applicability to study performed under grant and contract** Was the laboratory, contractor, or grantee informed that their services must be conducted in compliance with 40 CFR Part 792?				
§792.12	**Compliance statement**				
	(a) Was a compliance statement signed by the applicant?				
	the sponsor?				
	the study director?				
	(b) Was the compliance statement completed and submitted with the study report?				
	(c) Did the compliance statement include any statement of differences from the GLP regulations?				

TSCA GLP INSPECTION CHECKLIST
Data Audit Review
Revised 9/93

Laboratory: _____   Insp. Init.: _____   Date: _____

Comments (Please refer to subpart, section, or page numbers):

_____
_____
_____
_____
_____
_____
_____
_____
_____
_____
_____
_____
_____
_____
_____
_____
_____
_____
_____
_____
_____
_____
_____
_____
_____
_____
_____
_____
_____
_____
_____
_____

TSCA GLP INSPECTION CHECKLIST
Data Audit Review
Revised 9/93

Laboratory: _____ Insp. Init.: _____ Date: _____

FORM II — DATA AUDIT REVIEW (Continued)

SUBPART B — ORGANIZATION & PERSONNEL	YES	NO	N/A	REMARKS
§792.29 **Personnel**				
(a) Were training, education, and experience adequate?				
(b) Were training and experience records available?				
(c) Was the number of personnel adequate?				
§792.31 **Testing facility management**				
(a) Was a study director designated prior to study initiation?				
(b) Was the study director replaced during the course of the study?				
If so, was this done promptly?				
(c) Was a quality assurance unit in place?				
(d) Are personnel, resources, facilities, equipment, materials, and methodologies available for inspection?				
(e) Were deviations in the study communicated to the study director and corrective actions taken and documented?				
§792.33 **Study director**				
Did the study director have adequate education, training, and experience?				
Did the study director understand that his/her responsibilities included the following assurances:				
(a) The protocol, including any change, was approved and followed?				
(b) All experimental data were accurately recorded and verified?				
(c) Unforeseen circumstances were noted and corrective action taken and documented?				
(d) Test systems were as specified in the protocol?				
(e) All GLPs were followed?				
(f) All required data was transferred to the archives?				
§792.35 **Quality Assurance Unit**				
(a) Was a separate and independent QAU in place at the time of the study?				

TSCA GLP INSPECTION CHECKLIST
Data Audit Review
Revised 9/93

Laboratory: _____ Insp. Init.: _____ Date: _____

Comments (Please refer to subpart, section, or page numbers):

_____
_____
_____
_____
_____
_____
_____
_____
_____
_____
_____
_____
_____
_____
_____
_____
_____
_____
_____
_____
_____
_____
_____
_____
_____
_____
_____
_____
_____

TSCA GLP INSPECTION CHECKLIST
Data Audit Review
Revised 9/93

Laboratory: _____ Insp. Init.: _____ Date: _____

FORM II — DATA AUDIT REVIEW (Continued)

SUBPART B — ORGANIZATION & PERSONNEL	YES	NO	N/A	REMARKS
§792.35 (b) Did the QAU:				
(1) Maintain a complete copy of the master schedule indexed by test substance? (The required elements include the test substance, test system, nature of study, date initiated, current status, identity of sponsor, name of study director.)				
(2) Maintain copies of protocols?				
(3) Perform periodic QA inspections and maintain proper records of each inspection?				
(4) Periodically submit to management and study director written status reports on each study, noting any problems and corrective actions taken?				
(5) Keep dates indicating when management and the study director were notified of inspection findings?				
(6) Determine that no deviations were made without proper authorization and documentation?				
(c) Are the responsibilities and procedures, records, and indexing methods recorded in writing?				
(d) Were these procedures available for review?				

SUBPART C — FACILITIES	YES	NO	N/A	REMARKS
§792.51 Specimen and data storage facilities				
Is space provided for archives?				
Is access to the archives limited?				

SUBPART D — EQUIPMENT	YES	NO	N/A	REMARKS
§792.61 Equipment design				
Was the equipment used in the generation of data and facility environmental control appropriately designed and of adequate capacity to function according to protocol requirements?				
§792.63 Maintenance and calibration of equipment				
(a) Was equipment adequately inspected, maintained, and calibrated/standardized as required?				
(b) Did the SOPs adequately address the methods, materials, and schedules to be used in routine inspection, cleaning, maintenance, testing, and calibration/standardization of equipment, including action taken in case of a malfunction?				
(c) Were written records maintained of all inspection, maintenance, testing, and/or calibrating/standardization operations?				

TSCA GLP INSPECTION CHECKLIST
Data Audit Review
Revised 9/93

Laboratory: _____ Insp. Init.: _____ Date: _____

Comments (Please refer to subpart, section, or page numbers):

_____
_____
_____
_____
_____
_____
_____
_____
_____
_____
_____
_____
_____
_____
_____
_____
_____
_____
_____
_____
_____
_____
_____
_____
_____
_____
_____
_____
_____
_____
_____
_____
_____
_____

TSCA GLP INSPECTION CHECKLIST
Data Audit Review
Revised 9/93

Laboratory: _____ Insp. Init.: _____ Date: _____

FORM II — DATA AUDIT REVIEW (Continued)

SUBPART D — EQUIPMENT	YES	NO	N/A	REMARKS
§792.63(c) - Did these records describe whether the maintenance operations were routine and followed the SOPs?				
- Were written records kept of all **non-routine** repairs performed as a result of failure or malfunction?				
- Did the non-routine records document the nature of the defect, how and when the defect was discovered, and the remedial action taken in response?				
- Were the records signed or initialled and dated by the person making the entries?				

SUBPART E — TESTING FACILITIES OPERATIONS	YES	NO	N/A	REMARKS
§792.81 Standard Operating Procedures				
(a) Were written SOPs in place during the study adequate and available for review?				
- Are deviations from the SOP adequately documented in the raw data?				
- Were all significant changes properly authorized in writing by management?				
(b) Were written SOPs available for the following:				
(1) Test system area preparation?				
(2) Test system care?				
(3) Receipt, ID, storage, handling, mixing, and method of sampling of the test, control, and reference substances?				
(4) Test system observations?				
(5) Laboratory or other tests?				
(6) Handling of test systems found moribund or dead?				
(7) Necropsy of test systems or postmortem examination of test systems?				
(8) Collection and ID of specimens?				
(9) Histopathology?				
(10) Data handling, storage, and retrieval?				
(11) Maintenance and calibration of equipment?				
(12) Transfer, proper placement, and ID of test systems?				
(c) Is a historical file of SOPs and dates of revisions maintained?				

TSCA GLP INSPECTION CHECKLIST
Data Audit Review
Revised 9/93

Laboratory: _____ Insp. Init.: _____ Date: _____

Comments (Please refer to subpart, section, or page numbers):

_____
_____
_____
_____
_____
_____
_____
_____
_____
_____
_____
_____
_____
_____
_____
_____
_____
_____
_____
_____
_____
_____
_____
_____
_____
_____
_____
_____
_____
_____
_____
_____
_____

TSCA GLP INSPECTION CHECKLIST
Data Audit Review
Revised 9/93

Laboratory: _____ Insp. Init.: _____ Date: _____

FORM II — DATA AUDIT REVIEW (Continued)

SUBPART E — TESTING FACILITIES OPERATIONS	YES	NO	N/A	REMARKS
§792.83 Reagents and solutions				
Are records for reagents and solutions available that would indicate identity, concentration, storage requirements, and expiration date?				
§792.90 Animal and other test system care				
(a) Were SOPs for housing, feeding, handling, and care of test systems available?				
(b) At the initiation of the study, were test systems free of disease and appropriate for the study?				
- If test systems developed a disease or condition during the study, were test systems isolated?				
- Were test systems treated for the condition in such a manner that treatment did not interfere with the study?				
- Were the diagnosis, authorization of treatment, description of treatment, and dates of treatment documented in the raw data?				
(c) Were test systems needing to be removed from their housing units adequately identified (e.g., tattoo, color code, ear tag, ear punch, etc.)?				
- Were test system housing units adequately identified?				
(d) Were different species housed in separate rooms as necessary?				
- Were test systems of the same species, used for different studies, housed in separate rooms?				
- If the species were not housed in separate rooms, was adequate differentiation by space and identification made?				
(1) Were plants, invertebrate animals, and aquatic vertebrate animals used in multispecies tests, if housed in the same room, segregated to avoid mix-up or cross contamination?				
(e) Are records available indicating whether cages, racks, pens, enclosures, aquaria, holding tanks, ponds, growth chambers, and other holding, rearing, and breeding areas, and accessory equipment were cleaned and sanitized at appropriate intervals?				
(f) Were feed, soil, and water analyzed periodically for contaminants?				
- Was documentation maintained for these analyses?				
(g) Was the bedding used of a type that would not interfere with the conduct of the study?				
- Was the bedding changed as often as necessary?				

TSCA GLP INSPECTION CHECKLIST
Data Audit Review
Revised 9/93

Laboratory: _____ Insp. Init.: _____ Date: _____

Comments (Please refer to subpart, section, or page numbers):

TSCA GLP INSPECTION CHECKLIST
Data Audit Review
Revised 9/93

Laboratory: _____

Insp. Init.: _____ Date: _____

FORM II — DATA AUDIT REVIEW (Continued)

SUBPART E — TESTING FACILITIES OPERATIONS	YES	NO	N/A	REMARKS
§792.90 (h) If any pest control materials were used, was their use documented?				
- Were pest control materials used that would not interfere with the study?				
(i) Were test systems acclimated to the environmental conditions of the test?				

SUBPART F — TEST, CONTROL, AND REFERENCE SUBSTANCES

§792.105 Test, control, and reference substance characterization

(a) Were the substances characterized?	Test	Control	Reference	Documentation
- Identity				
- Strength				
- Purity				
- Stability				
- Uniformity				

SUBPART F — TEST, CONTROL, AND REFERENCE SUBSTANCES	YES	NO	N/A	REMARKS
§792.105(a) - Were methods of synthesis, fabrication, or derivation of the test, control, or reference substance documented and the location specified?				
- Was the location of documentation specified?				
(b) Were the solubility and/or stability of the test substance determined before the experiment start date?				
(c) Did each storage container for a test, control, or reference substance include the following information:				
- name, chemical abstracts service number (CAS) or code number?				
- batch number?				
- expiration date, if any?				
- storage conditions, if appropriate?				
- Were storage containers assigned to a particular test substance for the duration of the study?				
(d) For studies of more than 4 weeks experimental duration, were reserve samples from each batch of test, control, and reference substances retained for the period of time provided in §792.195?				
- Where are reserve samples archived?				
(e) Was the stability of the substance under the storage conditions at the test site known for all studies?				

TSCA GLP INSPECTION CHECKLIST
Data Audit Review
Revised 9/93

Laboratory: _____ Insp. Init.: _____ Date: _____

Comments (Please refer to subpart, section, or page numbers):

_____
_____
_____
_____
_____
_____
_____
_____
_____
_____
_____
_____
_____
_____
_____
_____
_____
_____
_____
_____
_____
_____
_____
_____
_____
_____
_____
_____
_____
_____
_____
_____

TSCA GLP INSPECTION CHECKLIST
Data Audit Review
Revised 9/93

Laboratory: _____ Insp. Init.: _____ Date: _____

FORM II — DATA AUDIT REVIEW (Continued)

SUBPART F — TEST, CONTROL, AND REFERENCE SUBSTANCES	YES	NO	N/A	REMARKS
§792.107 Test, control, and reference substance handling				
(a) Did an SOP covering handling of substances exist?				
(b) Were the substances stored according to the SOP?				
(c) Was distribution made so as to preclude the possibility of contamination, deterioration, or damage?				
(d) Was proper ID of substances maintained throughout the distribution process?				
(e) Was documentation maintained, including date and quantity of each receipt and distribution?				
§792.113 Mixtures of substances with carriers				
(a) Was appropriate analytical testing performed for each test, control, or reference substance:				
(1) To determine uniformity?				
- To determine, periodically, the concentration of the test, control, or reference substance in the mixture?				
(2) To determine solubility in the mixture, if necessary?				
- Was solubility testing done before the experimental start date?				
(3) To determine the stability in the mixture before the experimental start date or according to the SOP?				
(b) Was the expiration date shown on the mixture container, if necessary?				
(c) Was assurance made that the vehicle did not interfere with the integrity of the test?				

SUBPART G — PROTOCOL FOR AND CONDUCT OF A STUDY	YES	NO	N/A	REMARKS
§792.120 Protocol				
(d) Did the study have an approved written protocol indicating objectives and all methods?				
- Did each protocol contain at least the following:				
(1) A descriptive title and statement of purpose?				
(2) Identification of the test, control, and reference substance by name, CAS number, or code number?				
(3) Name and address of both sponsor and testing facility?				

Laboratory: _____ Insp. Init.: _____ Date: _____

Comments (Please refer to subpart, section, or page numbers):

_____
_____
_____
_____
_____
_____
_____
_____
_____
_____
_____
_____
_____
_____
_____
_____
_____
_____
_____
_____
_____
_____
_____
_____
_____
_____
_____
_____
_____
_____
_____
_____
_____

TSCA GLP INSPECTION CHECKLIST
Data Audit Review
Revised 9/93

Laboratory: _____   Insp. Init.: _____   Date: _____

FORM II — DATA AUDIT REVIEW (Continued)

SUBPART G — PROTOCOL FOR AND CONDUCT OF A STUDY			YES	NO	N/A	REMARKS
§792.120(a)	(4)	Proposed experimental start and termination dates?				
	(5)	Justification for selection of the test system?				
	(6)	Where applicable, the number, body weight, range, sex, source of supply, species, strain, substrain, and age of the test system?				
	(7)	Procedure for identification of the test system?				
	(8)	Description of the experimental design, including methods for the control of bias?				
	(9)	A description and/or identification of the:				
		- diet used in the study?				
		- solvents, emulsifiers and/or other materials used to solubilize or suspend the test, control, or reference substance before mixing with the carrier?				
		- specifications for acceptable levels of contaminants?				
	(10)	Route of administration and reason for its choice?				
	(11)	Dosage level in appropriate units and method and frequency of administration?				
	(12)	Type and frequency of tests, analyses, and measurements to be made?				
	(13)	The records to be maintained?				
	(14)	The date of approval of the protocol by the sponsor?				
		- The dated signature of the study director?				
	(15)	A statement of the proposed statistical method to be used?				
	(e)	Were all changes or revisions and reasons:				
		- documented?				
		- signed by the study director?				
		- dated?				
		- maintained with the protocol?				
§792.130		**Conduct of a study**				
	(a)	Was the study conducted in accordance with the protocol?				

TSCA GLP INSPECTION CHECKLIST
Data Audit Review
Revised 9/93

Laboratory: _____ Insp. Init.: _____ Date: _____

Comments (Please refer to subpart, section, or page numbers):

_____
_____
_____
_____
_____
_____
_____
_____
_____
_____
_____
_____
_____
_____
_____
_____
_____
_____
_____
_____
_____
_____
_____
_____
_____
_____
_____
_____
_____
_____
_____
_____
_____
_____

TSCA GLP INSPECTION CHECKLIST
Data Audit Review
Revised 9/93

Laboratory: _____ Insp. Init.: _____ Date: _____

FORM II — DATA AUDIT REVIEW (Continued)

SUBPART G — PROTOCOL FOR AND CONDUCT OF A STUDY	YES	NO	N/A	REMARKS
§792.130 (b) Were the test systems monitored in conformity with the protocol?				
(c) Were the specimens identified by:				
- test system?				
- study?				
- nature of collection?				
- date of collection?				
- Was the specimen information either on the container or accompanying the specimen described in a manner that precludes error?				
(d) If applicable, were the gross necropsy observations available to the pathologist for the histopathological exam?				
(e) Were all data recorded promptly and legibly in ink?				
- Were all data entries (non-automated) signed (or initialed) and dated on the day of entry?				
- Were changes in entries made so as not to obscure the original entry?				
- Were reasons given for changes?				
- Were changes identified and dated?				
- For automated data, was the individual responsible for direct data input identified at the time of data input?				
§792.135 Physical and chemical characterizations studies				
(a) Were all provisions of the GLP standards applied to physical and chemical characterization studies designed to determine stability, solubility, octanol water partition coefficient, volatility, and persistence of test, control, or reference substances?				
SUBPART J — RECORDS AND REPORTS	YES	NO	N/A	REMARKS
§792.185 Reporting of study results				
(a) Was a final report prepared to contain at least the following?				
(1) Name and address of the facility performing the study?				
- the dates on which the study was initiated, completed, terminated, or discontinued?				

TSCA GLP INSPECTION CHECKLIST
Data Audit Review
Revised 9/93

Laboratory: _____ Insp. Init.: _____ Date: _____

Comments (Please refer to subpart, section, or page numbers):

_____
_____
_____
_____
_____
_____
_____
_____
_____
_____
_____
_____
_____
_____
_____
_____
_____
_____
_____
_____
_____
_____
_____
_____
_____
_____
_____
_____
_____
_____
_____
_____

TSCA GLP INSPECTION CHECKLIST
Data Audit Review
Revised 9/93

Laboratory: _____ Insp. Init.: _____ Date: _____

FORM II — DATA AUDIT REVIEW (Continued)

SUBPART J — RECORDS AND REPORTS			YES	NO	N/A	REMARKS
§792.185(a)	(2)	The objectives and procedures as stated in the approved protocol?				
	(3)	The statistical methods employed?				
	(4)	The test, control, and reference substance identified by name, CAS number or code number, strength, purity, and composition?				
	(5)	Stability and, if needed, solubility, of the substances under conditions of administration?				
	(6)	A description of the methods used?				
	(7)	A description of the test system used?				
		- Where applicable, the number of animals or other test systems used, sex, body weight range, source of supply, species, strain and substrain, age, and procedures used for ID?				
	(8)	A description of the dosage, dosage regimen, route of administration, and duration?				
	(9)	A description of all of the circumstances that may have affected the quality or integrity of the data?				
	(10)	The name of the study director?				
		- The names of other scientists, professionals, and supervisory personnel?				
	(11)	A description of the transformations, calculations, or operations performed on the data?				
		- A summary and analysis of the data?				
		- A statement of conclusions drawn from the data?				
	(12)	Signed and dated reports of each of the individual scientists or other professionals involved in the study, including each person who conducted an analysis or evaluation of data or specimens?				
	(13)	The locations where all specimens, raw data, and the final report are to be stored?				
	(14)	A QAU statement prepared and signed as specified in §792.35(b)(7)?				
	(b)	Was the final report signed and dated by the study director?				
	(c)	Were corrections or additions to the final report in the form of an amendment by the study director?				

TSCA GLP INSPECTION CHECKLIST
Data Audit Review
Revised 9/93

Laboratory: _____ Insp. Init.: _____ Date: _____

Comments (Please refer to subpart, section, or page numbers):

_____
_____
_____
_____
_____
_____
_____
_____
_____
_____
_____
_____
_____
_____
_____
_____
_____
_____
_____
_____
_____
_____
_____
_____
_____
_____
_____
_____
_____
_____
_____
_____
_____
_____
_____

TSCA GLP INSPECTION CHECKLIST
Data Audit Review
Revised 9/93

Laboratory: _____
Insp. Init.: _____ Date: _____

FORM II — DATA AUDIT REVIEW (Continued)

SUBPART J — RECORDS AND REPORTS		YES	NO	N/A	REMARKS
§792.185(c)	- Were the amendments clearly identified with:				
	- reasons for change?				
	- date?				
	- signature of person responsible?				
	(d) Is a copy of the final report with amendments maintained by the sponsor and the test facility?				
§792.190	Storage and retrieval of records and data				
	(a) Where are the raw data for the study archived in compliance with this section?				
	(b) Were all raw data, documentation, records, protocols, specimens, and final reports retained which were generated as a result of a study?				
	- Were all correspondence and other documents relating to interpretation and evaluation of data, other than those contained in the final report, retained?				
	(c) Are archives provided for orderly storage and expedient retrieval of all raw data, documentation, protocols, specimens, and interim and final reports?				
	- Are the conditions of the storage area appropriate to minimize deterioration in accordance with the time period of their retention and the nature of the documents or specimens?				
	(d) Is an individual responsible for the archives?				
	(e) Is it specified that only authorized personnel have access to the archives?				
	(f) Is the material retained in the archives indexed for rapid retrieval?				
§792.195	Retention of records				
	(b)				
	(1) Were records retained for at least ten years following the effective date of an applicable final test rule?				
	(2) Did the sponsor negotiate a testing agreement?				
	If Yes, were the data retained for at least ten years following the publication date of the acceptance of the negotiated agreement?				
	(3) In the case of testing submitted under TSCA Section 5, were records retained for at least five years following submission to EPA?				

TSCA GLP INSPECTION CHECKLIST
Data Audit Review
Revised 9/93

Laboratory: _____ Insp. Init.: _____ Date: _____

Comments (Please refer to subpart, section, or page numbers):

_____
_____
_____
_____
_____
_____
_____
_____
_____
_____
_____
_____
_____
_____
_____
_____
_____
_____
_____
_____
_____
_____
_____
_____
_____
_____
_____
_____
_____
_____
_____
_____

TSCA GLP INSPECTION CHECKLIST
Data Audit Review
Revised 9/93

Laboratory: _____ Insp. Init.: _____ Date: _____

FORM II — DATA AUDIT REVIEW (Continued)

SUBPART J — RECORDS AND REPORTS	YES	NO	N/A	REMARKS
§792.195 (d) Are wet specimens, samples of test, control, or reference substances, and specially prepared material that are relatively fragile and differ markedly in stability and quality during storage retained only as long as the quality of the preparation affords evaluation?				
(e) Were the master schedule sheet, copies of protocols, and records of quality assurance inspections, as required by 792.35(c), maintained by the QAU as an easily-accessible system of records for the period of time specified in questions 1 or 2 of this section?				
(f) Were provisions in place to retain summaries of training, experience, and job descriptions, required to be maintained by 792.29(b), as well as all other testing facility employment records for the length of time specified in questions 1 or 2 of this section?				
(g) Were provisions in place to retain records and reports of the maintenance, calibration, and inspection of equipment, as required by 792.63(b) and (c), for the length of time specified in questions 1 or 2 of this section?				
(h) Were provisions in place to retain records required by this part either as original records or as true copies such as photocopies, microfilm, microfiche, or other accurate reproductions of the original records?				

TSCA GLP INSPECTION CHECKLIST
Data Audit Review
Revised 9/93

About Government Institutes

Government Institutes, Inc. was founded in 1973 to provide continuing education and practical information for your professional development. Specializing in environmental, health and safety concerns, we recognize that you face unique challenges presented by the ever-increasing number of new laws and regulations and the rapid evolution of new technologies, methods and markets.

Our information and continuing education efforts include a Videotape Distribution Service, over 140 courses held nation-wide throughout the year, and over 150 publications, making us the world's largest publisher in these areas.

Government Institutes, Inc.
4 Research Place, Suite 200
Rockville, MD 20850
(301) 921-2300

Other related books published by Government Institutes:

TSCA Handbook, 2nd Edition — The national law firm of McKenna & Cuneo details existing chemical regulation under TSCA; EPA's program for evaluating and regulating new chemical substances; PMN preparations and follow through; civil and criminal liability; inspections and audits; required testing of chemical substances and mixtures; exemptions from PMN requirements; the TSCA Chemical Inventory; reporting and retention of information; and special obligations of importers/exporters. Contains charts, figures, tables, and multiple indexes. ISBN: 0-86587-791-2 *Softcover/490 pages/ Nov '89* **$95** ($114)

TSCA Inspection Guidance — New updated version of the manual developed by EPA to support its field inspectors in conducting compliance monitoring inspections of the manufacturing, processing, distribution, use, and disposal of chemicals. Contains detailed inspection guidance for ensuring compliance with TSCA Sec. 5 (DMN) and Sec. 8 (Recordkeeping & Reporting)—as well as Sec. 4 (Testing), Sec. 12 (Exports), Sec. 13 (Imports), and a review of the current biotechnology program under TSCA. ISBN: 0-86587-358-5 *Softcover/ 400 pages/Nov '93* **$85** ($102)

TSCA Confidential Business Information Security Manual — Contains the stringent procedures to be followed by EPA personnel, other Federal agencies, and contractors in handling information, submitted or collected under TSCA, that is claimed by companies as confidential business information. These confidentiality procedures are designed to provide companies their right to declare inspection data confidential and to ensure secure handling of this information at each stage of a TSCA inspection. ISBN: 0-86587-363-1 *Softcover/154 pages/Oct '93* **$69** ($83)

Environmental Law Handbook, 12th Edition — The recognized authority in the field, this invaluable text, written by nationally-recognized legal experts, provides practical and current information on all major environmental areas. ISBN: 0-86587-350-X *Hardcover/670 pages/Apr '93* **$72** ($86)

Directory of Environmental Information Sources, 4th Edition — Details hard-to-find Federal Government Resources; State Government Resources; Professional, Scientific, and Trade Organizations; Newsletters, Magazines, and Periodicals; and Databases. ISBN: 0-86587-326-7 *Softcover/350 pages/Nov '92* **$78** ($94)

Environmental Audits, 6th Edition — Details how to begin and manage a successful audit program for your facility. Use these checklists and sample procedures to identify your problem areas. ISBN: 0-86587-776-9 *Softcover/592 pages/Nov '89* **$79** ($95)

Call the above number for our current book/video catalog and course schedule.

PRICES ARE SUBJECT TO CHANGE. U.S. PRICES ARE IN BOLD FACE. (OUTSIDE U.S. PRICES ARE IN PARENTHESES.)